IIW Collection

International Institute of Welding, Genova, Italy

The IIW Collection of Books is authored by experts from the 59 countries participating in the work of the 23 Technical Working Units of the International Institute of Welding, recognized as the largest worldwide network for welding and allied joining technologies.

The IIW's Mission is to operate as the global body for the science and application of joining technology, providing a forum for networking and knowledge exchange among scientists, researchers and industry.

Published books, Best Practices, Recommendations or Guidelines are the outcome of collaborative work and technical discussions-they are truly validated by the IIW groups of experts in joining, cutting and surface treatment of metallic and non-metallic materials by such processes as welding, brazing, soldering, thermal cutting, thermal spraying, adhesive bonding and microjoining. IIW work also embraces allied fields including quality assurance, non-destructive testing, standardization, inspection, health and safety, education, training, qualification, design and fabrication.

Guy Parmentier · Michel Huther · Isabel Huther · Fabien Lefebvre

Best Practice Guideline for Statistical Analyses of Fatigue Results

INTERNATIONAL INSTITUTE OF WELDING
A world of joining experience

Springer

Guy Parmentier
Bureau Veritas Marine and Offshore
Puteaux, France

Michel Huther
Bureau Veritas Marine and Offshore
Puteaux, France

Isabel Huther
CETIM
Senlis, France

Fabien Lefebvre
CETIM
Senlis, France

ISSN 2365-435X ISSN 2365-4368 (electronic)
IIW Collection
ISBN 978-3-031-23569-6 ISBN 978-3-031-23570-2 (eBook)
https://doi.org/10.1007/978-3-031-23570-2

This Springer imprint is published by the registered company Springer Nature Switzerland AG
The registered company address is: Gewerbestrasse 11, 6330 Cham, Switzerland

Preface

This guide is the joining together of the following IIW documents:

- Best practice guide on statistical analysis of fatigue data (doc-XIII-2138-06):

C. R. A. Schneider	TWI (Cambridge—UK)
S. J. Maddox	TWI (Cambridge—UK)

- Guidance for the application of the best practice guide on statistical analysis of fatigue—working sheets (doc XIII-WG1-188-17):

G. Parmentier	Bureau Veritas (Paris La Défense—France)
M. Huther	Bureau Veritas (Paris La Défense—France)
A. Galtier	Arcelor Research (Maizières les Metz—France)
I. Huther	CETIM (Senlis—France)
G. Marquis	Department of Mechanical Engineering, School of Engineering, Aalto University (Espoo—Finland)

Recommendations for analysing fatigue data are available, but they do not deal with all the statistical treatments that may be required to utilise fatigue test results, and none of them offers specific guidelines for analysing fatigue data obtained from tests on welded specimens.

The subject of the present guide is to provide to engineers a comprehensive guidance for the use of sound statistical methods and for the evaluation of fatigue data of welded components and structures obtained under constant amplitude loading and used to produce SN curves.

For an easy use, working sheets are provided to assist in the proper statistical assessment of experimental fatigue data concerning practical problems giving the procedure and a numerical application as illustration.

Puteaux, France	Guy Parmentier
Puteaux, France	Michel Huther
Senlis, France	Isabel Huther
Senlis, France	Fabien Lefebvre

Contents

Chapter 1
Best Practice on Statistical Analysis of Fatigue Data

1.1 Introduction

Fatigue testing is the main basis of the relationship between the fatigue resistance of a given material, component or structural detail and cyclic loading. The results of such fatigue endurance tests are plotted on graphs relating applied loading (force, stress, strain, etc.) and the number of cycles to failure. Since test specimens and testing conditions are never identical, the resulting data are invariably scattered. Consequently, some judgement is required when using them to establish the required relationship. Statistical methods are available to assist in this analysis of fatigue test data, and indeed some recommendations on their use for analysing fatigue data are available [1, 2]. However, they do not deal with all the statistical analyses that may be required to utilise fatigue test results and none of them offers specific guidelines for analysing fatigue data obtained from tests on welded specimens. With the increasing use of fatigue testing to supplement design rules, an approach that is now encouraged in some Standards [3, 4, 25], there is a need for comprehensive guidance on the statistical analysis of fatigue test results.

This is the subject of the present best practice guide. At this stage, the focus is on fatigue endurance test results obtained under constant amplitude loading, as used to produce *S–N* curves. Thus, the loading is expressed as a stress range, *S*, and the fatigue resistance is expressed as the number of cycles, *N*, that can be endured by the test specimen. In general, however, the same methods can be applied to fatigue endurance test results expressed using any measure of the loading (e.g. force, strain) and results obtained under variable amplitude loading. They can also be used to analyse fatigue crack propagation data, where the loading is expressed as the stress intensity factor range, ΔK, and the fatigue resistance is expressed as the rate of crack propagation da/dN. Since the analyses are concerned purely with the experimental data, they are independent of the material tested.

G. Parmentier et al., *Best Practice Guideline for Statistical Analyses of Fatigue Results*, IIW Collection, https://doi.org/10.1007/978-3-031-23570-2_1

1.2 Assumptions

1.2.1 Form of S–N Curve

(a) There is an underlying linear relationship between log S and log N of the form:

$$\log N = \log A - m \log S \tag{1.1}$$

where m is the slope and log A is the intercept. This can be rewritten in a form that is commonly used to describe S–N curves in design rules:

$$S^{m} N = A \tag{1.2}$$

Note that, in practice, this assumption will only hold true between certain limits on S, as illustrated in Fig. 1.1. The lower limit on S is determined by the fatigue endurance limit (or just "fatigue limit"), the stress range below which fatigue failure will not occur. In practice this is usually chosen on the basis of the endurance that can be achieved without any evidence of fatigue cracking, typically between $N = 2 \times 10^6$ and 10^7 cycles. The upper limit on S is dependent on the static strength of the test specimen but is commonly taken to be the maximum allowable static design stress [5].

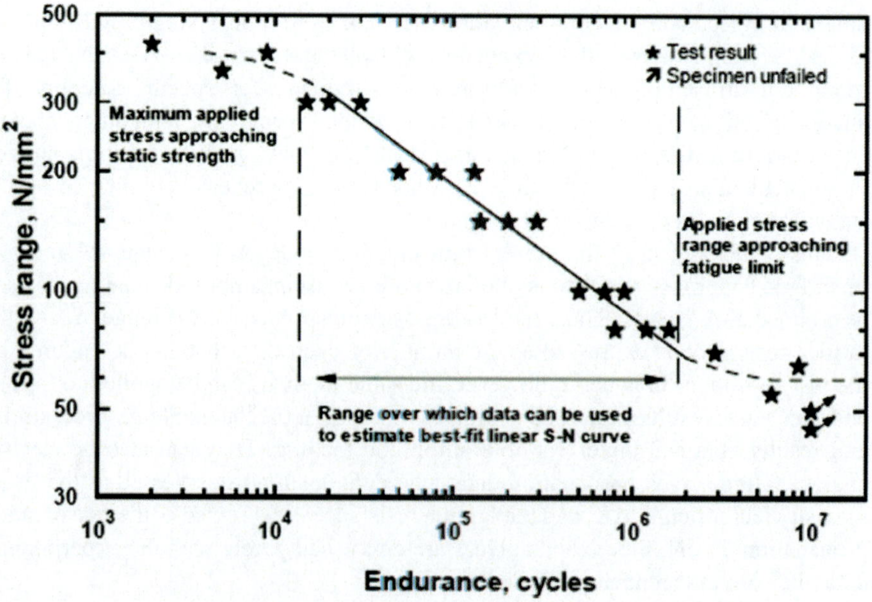

Fig. 1.1 Typical fatigue endurance test data illustrating deviations from linear S–N curve

However, the linear relationship between applied strain range and fatigue life for data obtained under strain control extends to much higher pseudo-elastic stress (i.e. strain × elastic modulus) levels.

(b) The fatigue life *N* for a given stress range *S* is log-normally distributed.
(c) The standard deviation of log *N* does not vary with *S*.
(d) Each test result is statistically independent of the others.

These assumptions are rarely challenged in practice. But, if there is any reason to doubt their validity, there are statistical tests available that can help to identify departures from these assumptions. Some of these tests are listed below.

1.2.2 Tests for Linearity of Relationship Between log S and log N

Many statistical software packages (e.g. MINITAB [6]) include statistical tests for lack of fit to the linear regression model. If the data include replicates (i.e. more than one observation of *N* for some of the values of *S*), then lack of fit can be tested by means of Analysis of Variance (ANOVA). This technique uses an *F*-test to compare the variance within the groups of replicates with the variance of the data about the fitted straight line. Lack of fit is indicated if the variance about the line is significantly larger than the variance within replicates (see p. 230 of Cooper [7] for a worked example). This test is also called a "pure error" lack of fit test [6].

Another common method of testing for linearity in a relationship is to fit a polynomial (typically a quadratic or cubic) in log *S* to the data. Polynomial regression is available within most statistical software packages (e.g. MINITAB [6]). The ANOVA method can then be used to test whether the quadratic/cubic regression components are statistically significant (see Ferguson [8] for a worked example).

Box and Tidwell [9] describe a less well-known approach, which is to add a term of the form (log *S*)ln(log *S*) to the usual linear regression model. If the coefficient in this variable is significant, then this can be taken as evidence of nonlinearity.

1.2.3 Tests that N is Log-Normally Distributed

The simplest check is an "eyeball" assessment of whether a normal probability plot of the departures (or "residuals") of log *N* from the regression line of log *S* versus log *N* follows a linear trend. This can be done either by using standard statistical software, or by plotting the residuals on normal probability paper.

There is also a wide variety of formal statistic-based tests of normality, many of which are implemented in statistical software packages [10, 11].

1.2.4 Tests of Homogeneity of Standard Deviation of log N with Respect to S

This assumption is most easily checked by simply examining a plot of the "residuals" from the regression versus log *S*. The assessment can be backed up by partitioning the residuals into appropriate groups and applying either Bartlett's test [7], if log *N* is believed to be normally distributed, or otherwise Levene's test [6].

1.2.5 Tests of Statistical Independence of Test Results

This assumption is difficult to check, in practice. A good starting point is to examine plots of the "residuals" from the regression against both log *S* and against the order in which the results were collected (in case there is some time dependence). There should not be any recognisable patterns in the residuals in either of these plots. If there are, the data can be grouped accordingly, and variations in the mean level can then be tested using Analysis of Variance. Any inhomogeneity in the standard deviations of the groups can be tested as in Sect. 1.2.4.

1.3 Fitting a *S–N* Curve

In their simplest form, *S–N* data comprise n data points (log S_i, log N_i), where S_i is the stress range and N_i is the endurance in cycles. This endurance is either the number of cycles to failure (or some predetermined criterion, such as the attainment of a particular size of fatigue crack) or the number of cycles endured without failure. Figure 1.2 shows an example of such data, together with some fitted *S–N* curves [12].

Special attention is drawn to the fact that fatigue test results are traditionally plotted with log *S* as the *y*-axis and log *N* as the *x*-axis. The standard approach in curve fitting is to assume that the parameter plotted on the *x*-axis is the independent variable and that plotted on the *y*-axis is the dependent variable. However, the opposite is the case with fatigue data presented in the traditional way. Consequently, care is needed to ensure that log *N* is treated as the **dependent** variable.

Considering only the results from specimens that failed, the intercept log *A* and slope *m* of the "best fit" line through the data (called the "mean" line in Fig. 1.2) are estimated by ordinary linear regression, as described by Gurney and Maddox [5]. The method usually used to estimate the slope and intercept coefficients is called "least squares estimation". This method is based on choosing those values of the coefficients that minimise the sum of the squared deviations (or "residuals") of the

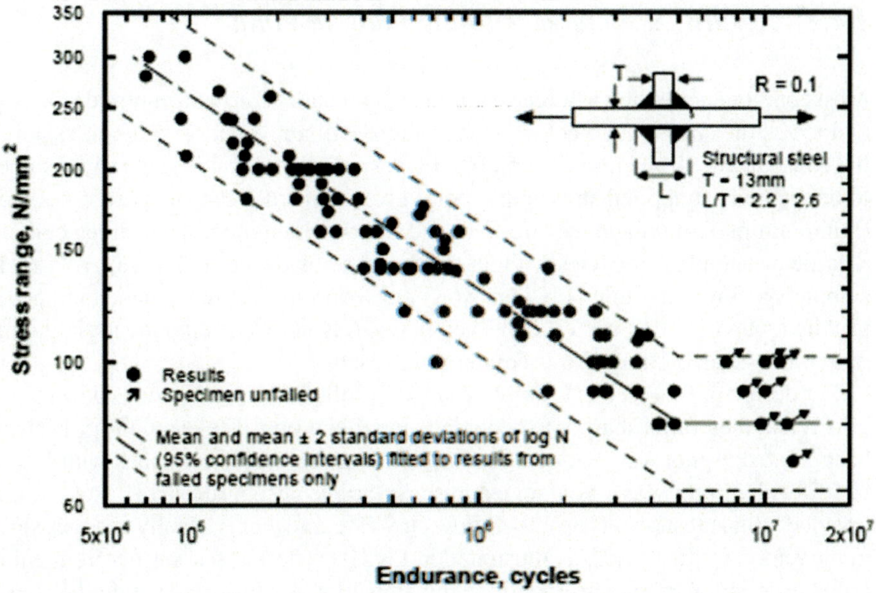

Fig. 1.2 Example of *S–N* data (Maddox [12])

observed values of log N_i from those predicted by the fitted line. TWI originally used such data and this method of analysis to derive the fatigue design rules for welded steel structures that have since formed the basis of most fatigue design rules in the world [5, 13].

1.4 Treatment of Results Where No Failure Has Occurred

1.4.1 Introduction

In Sect. 1.3, it was assumed that each specimen under test yielded an exact failure endurance. However, there are circumstances in which results are obtained from specimens, or parts of specimens, that have not failed. Such results, which are plotted in Figs. 1.1 and 1.2 as "specimen unfailed", are often termed "run-outs". Depending on the circumstances, it may or may not be possible to use the results from unfailed specimens in the statistical analysis of the data. Indeed, even nearby results from specimens that did fail may need to be excluded from that analysis.

1.4.2 Results Associated with the Fatigue Limit

Most components exhibit a fatigue endurance limit under constant amplitude loading, defined as the stress range below which failure will not occur. In order to establish the constant amplitude fatigue limit (CAFL) experimentally, it is generally assumed to be the highest applied stress range for a given applied mean stress or stress ratio (minimum/maximum applied stress) at and below which the test specimen endures a particular number of cycles without showing any evidence of fatigue cracking. In smooth specimens, with no obvious stress concentration features, that endurance is usually around 2×10^6 cycles. However, in severely notched components, including most weld details, 10^7 cycles is commonly chosen.

For design purposes, it is usually assumed that the *S–N* curve extends down to the CAFL and then turns sharply to become a horizontal line. The data in Fig. 1.2 have been treated in this way, with the assumption that the CAFL coincides with $N = 5 \times 10^6$ cycles in this case, as assumed in some fatigue design rules [13]. However, in practice fatigue test results usually follow an *S–N* curve that gradually changes slope in the region of the CAFL, as illustrated in Fig. 1.1. Clearly, test results from either failed or unfailed specimens that lie in this transition region approaching the CAFL should not be combined with those obtained at higher stresses when estimating the best fit linear *S–N* curve. Some judgement is needed when deciding which results fall into this category but, as a guide, any from notched or welded specimens that give $N < 2 \times 10^6$ cycles could be included, or $N < 10^6$ cycles in the case of results obtained from smooth specimens. A test for linearity (see Sect. 1.2.2) could be used to confirm the choice.

Depending on the circumstances, it may be necessary to model the *S–N* curve more precisely and include the transition regions at high and low applied stresses shown in Fig. 1.1. In such a case, the data are no longer assumed to fit a linear log *S* versus log *N* relationship, but one that corresponds to an S-shaped curve, such as given by Bastenaire [14]:

$$N = \frac{A}{S - E} \exp\left[-\left(\frac{S - E}{B}\right)^{C}\right] \tag{1.3}$$

where *A*, *B*, *C* and *E* are constants to be estimated from the data.

1.4.3 Fatigue Tests Stopped Before Failure

Two other situations in which fatigue test results refer to unfailed specimens provide information that can be used in the estimation of the best fit *S–N* curve. In both cases, they are situations in which fatigue failure would have occurred eventually if testing had continued. Thus, results that lie in the transition region approaching the CAFL discussed in Sect. 1.4.2 are excluded. The two situations are:

(a) The test is stopped deliberately, perhaps because of time constraints.
(b) The test specimen contains more than one site for potential fatigue failure and fails from just one of them. At this stage, the remaining sites are only partly through their fatigue lives.

This second case was the situation in the welded specimens that gave the results in Fig. 1.2. Failure occurred as a result of fatigue cracking from the weld toes on one side of the attachment leaving the other side intact. Thus, their test results could have been included in Fig. 1.2 as "specimen unfailed". In other circumstances, a test will generally be stopped even if failure occurs at a completely unrelated location. In all these cases, it is clearly desirable to infer as much as possible from the locations where failure has not occurred (so-called censored data), as well as those where it has (called "exact" data).

1.4.3.1 Maximum Likelihood Method

The maximum likelihood method provides an appropriate tool for solving the general problem of estimating the "best fit" line through censored test data [15, 16]. In a quite fundamental sense, the maximum likelihood method provides estimates for the slope and gradient coefficients that maximise the likelihood of obtaining the observed data. Thus, the resulting estimates are those that agree most closely with the observed data. In the special case of exact data, the maximum likelihood method leads to the least squares function on which linear regression is generally based [15]. In the general case, numerical iteration is required to derive maximum likelihood estimates. Fortunately, linear regression of censored data has general application in the field of reliability analysis, and so many statistical software packages can perform the required calculations within a few seconds on a modern PC.

1.4.3.2 Alternative Method Based on Extreme Value Statistics

A special case of censored data can arise when testing a number of specimens, each of which contains the same number M of nominally identical welds and any of which might fail first. Maddox [17] (appended to this report for convenience) shows that, if each specimen is tested until it fails at exactly one of the M potential locations, then the S–N curve for a single weld can be established using least squares estimation, together with the tabulated extreme value statistics for the normal distribution. Although this approach is less flexible than maximum likelihood estimation, it can be performed and/or verified by hand calculation. It should also be noted that, in this case, least squares estimation is no longer equivalent to maximum likelihood estimation (because extreme value statistics for the normal distribution are not themselves distributed normally).

1.5 Establishing a Design (or Characteristic) Curve

1.5.1 Prediction Limits

For design purposes, it is necessary to establish limits between which a given proportion (typically 95%) of the data lies. These bounds are often termed "confidence limits" [5]. In this Guide, the term "prediction limits" is used instead, to avoid confusion with the confidence limits on the coefficients of the regression line.

In the case of exact data, considered in Sect. 1.3, the prediction limits at stress range S can be expressed explicitly, in the form [7]:

$$\log N_{p\%}^{\pm} = (\log A + m \log S) \pm t\hat{\sigma}\sqrt{1 + \frac{1}{n} + \frac{\left(\log S - \overline{\log S}\right)^2}{\sum_{i=1}^{n}\left(\log S_i - \overline{\log S}\right)^2}} \tag{1.4}$$

where

$\log A$ and m are the coefficients of the regression line through the n data points ($\log S_i$, $\log N_i$), as in Sect. 1.3.

$\overline{\log S}$ is the mean of the n values of $\log S_i$.

t is the appropriate percentage point of Student's t distribution, with f degrees of freedom.

$\hat{\sigma}^2$ is the best estimate of the variance of the data about the regression line, which is equal to the sum of squared residuals divided by the number of degrees of freedom f.

f is equal to $n - 2$ in the case where the two coefficients of the regression line have both been estimated from the data.

In general, these prediction limits are hyperbolae, rather than straight lines, which are closest to the regression line at the mean log-stress value $\overline{\log S}$. However, it is often assumed that design curves will only be applied to values of $\log S$ that are not far removed from the mean value $\overline{\log S}$ (see Gurney and Maddox [5] for a justification of this assumption). In this case, the third (final) term under the square root of Eq. (1.4) can be ignored, and the resulting prediction limits are parallel to the regression line.

Furthermore, as the sample size increases, the second term under the square root (i.e. $1/n$) becomes negligible; Gurney and Maddox [5] ignored this term for sample sizes larger than 20, which incurs an error of at most 2% in the width of the resulting prediction interval. A closely related situation is where the slope m is chosen to take a fixed value, e.g. $m = 3$ is usually chosen for welded steel joints for which the fatigue life is dominated by crack growth [5, 13]. In this case, the expression under the square root is exactly equal to one, and the number of degrees of freedom f should be increased by one (from $n - 2$ to $n - 1$). Note that this latter approach is generally recommended whenever the sample size is less than 10 [4].

As the sample size becomes even larger, the percentage points of the t distribution approach those of the normal distribution, so that approximate two-sided 95% prediction limits are, for example, given by substituting $t = 2$. IIW recommendations [4] indicate that this approximation can be used for sample sizes larger than 40 (again incurring an error of at most 2% in the width of the resulting prediction interval).

The two-sided 95% prediction limits are symmetrical, so there is a confidence of 97.5% of exceeding the lower limit $\log N_{95\%}$. This lower one-sided 97.5% prediction limit forms the basis of the most widely used fatigue design curves [13] (e.g. BS 5400, BS 7608, BS PD 5500, HSE Offshore Guidance, IACS Classification Society Rules).

1.5.2 Tolerance Limits

For sample sizes smaller than 40, the IIW document [4] suggests an alternative methodology for establishing a characteristic curve, based on estimating confidence limits on the prediction limits. Such limits are called "tolerance limits" [18], and they are further discussed in Sect. 1.10. The use of tolerance limits rather than prediction limits would yield a more conservative design curve, and tolerance limits have the advantage that they explicitly allow for uncertainty in estimates of population statistics (e.g. standard deviation) from a small sample. However, their use for design purposes would have the following disadvantages:

- They are inherently more complicated, and therefore there is an increased risk of misinterpretation.
- They are harder to implement (within a spreadsheet, for instance).
- The extra conservatism may not be warranted for all applications.
- Tolerance limits are more sensitive to deviations from the assumed normality than prediction limits.
- Their use represents a fundamental change to previous practice, so there is a risk of incompatibility with current design rules.

Therefore, it is not always appropriate to base a design curve on a tolerance limit. Tolerance limits are, nevertheless, a valuable tool for studying the sensitivity of a design curve to the size of the sample on which it is based (see Sect. 1.10). It is therefore recommended that tolerance limits be used as a means of justifying design curves that are based on small samples, especially for critical applications.

1.5.3 Results Where No Failure Has Occurred

For censored data, statistical software can be used to estimate prediction limits (or tolerance limits). Alternatively, where the approach of Sect. 1.4.3.2 applies, approximate prediction limits can be derived from the set of failed specimens, in which case they take the same form as Eq. (1.4).

1.6 Predicting Fatigue Life

1.6.1 Individual Weld

The mean fatigue life of an individual weld/sample is given by the "best fit" *S–N* curve (Sects. 1.3 and 1.4). The lower one-sided *P*% prediction limit N_{PL}^{-} is the best estimate of the fatigue life that will be exceeded by a given proportion *P*% of such weld details (Sect. 1.5).

1.6.2 Structure Containing Many Welds

Assuming there is no redundancy in the structure, the fatigue life of a structure containing *M* identical welds, any of which might fail, is given by the minimum of the *M* individual fatigue lives, thus:

$$N_{\min} = \min\{N_1, N_2, \ldots, N_M\} \tag{1.5}$$

where the individual fatigue lives N_1, N_2, ..., N_M are identically distributed.

This fatigue life will, on average, be lower than that of an individual weld. The mean fatigue life of the overall structure can be estimated (at least approximately) from the extreme value statistics of the normal distribution, as described by Maddox (see Appendix) [17]. However, prediction limits on the fatigue life of the structure are best obtained from the relationship:

$$\Pr(N_{\min} > N_{PL}^{-}) = [\Pr(N_1 > N_{PL}^{-})]^M \tag{1.6}$$

Thus, if N_{PL}^{-} is the lower one-sided *P*% prediction limit on the fatigue life of a single weld, then it is also equal to the lower one-sided *Q*% prediction limit on the fatigue life of the overall structure as long as $Q\% = [P\%]^M$. Hence, the lower one-sided *Q*% prediction limit on the fatigue life of the structure is equal to the $[Q\%]^{1/M}$ one-sided prediction limit for a single weld (which can be evaluated as in Sect. 1.5).

1.7 Justifying the Use of a Given Design Curve from a New Data Set

1.7.1 Problem

To validate the use of a particular design class on the basis of a limited number *n* of new fatigue tests it is assumed that the mean *S–N* curve for the particular design category (e.g. BS 7608 Class D) being validated, $S^m N = AD$, and the corresponding standard deviation of log N, σ, are known.

1.7.2 Approach

Initially the assumption is made that the new test results form part of the same population as that used to determine the design S–N curve (this is called the null hypothesis). Then, hypothesis testing is used to show that, under this assumption, it is very unlikely (at a specified significance level) that the new results would give such long fatigue lives. This is the basis for regarding the null hypothesis as implausible, and for accepting the alternative hypothesis that the new results actually belong to a population having longer fatigue lives than the main database.

1.7.3 Assumptions

In addition to the assumptions of Sect. 1.2, the analysis of this section depends on one or two extra assumptions concerning the compatibility of the new data set with the design curve. As for earlier assumptions, there are standard statistical tests available that can help to identify departures from the assumptions; these tests are also identified below.

(a) The slope of the mean S–N curve for the new test results is the same as the slope m of the design curve. This assumption is needed where a S–N curve is assumed for the new test results, i.e. only in Sect. 1.7.5.3. In these cases, TWI recommends the following test be routinely applied.

This assumption can be tested, at a given significance level $\alpha\%$ (e.g. the 5% level), by checking that m falls within the two-sided $(100 - \alpha)$ % confidence interval on the slope *mtest* of a regression line fitted through the new results. In the case of exact data, considered in Sect. 1.3, this confidence interval is given by [7]:

$$m_{\text{test},P\%}^{\pm} = m_{\text{test}} \pm t\hat{\sigma}\sqrt{\frac{1}{\sum_{i=1}^{n}\left(\log S_i - \overline{\log S}\right)^2}} \tag{1.7}$$

where

the new data points are given by $(\log S_i, \log N_i)$, for $i = 1, \ldots, n$.
$\overline{\log S}$ is the mean of the n values of $\log S_i$.
t is the two-sided $P\%$ percentage point (where $P = 100 - \alpha$) of Student's t distribution, with $n - 2$ degrees of freedom.

$\hat{\sigma}^2$ is the best estimate of the variance of the data about the regression line (as in Sect. 1.5.1).

(b) The standard deviation σ_{test} of log N about the mean S–N curve (having the assumed fixed slope m) is the same for the new tests as for the main database.

Under this assumption, the ratio $f\sigma_{\text{test}}^2/\sigma^2$ follows a χ^2 distribution (for exact data), with f degrees of freedom, where f is equal to n minus the number of coefficients estimated from the new data [19] ($f = n - 1$, usually). The assumption can thus be tested by reference to tabulated percentage points of the χ^2 distribution [18] (these are also commonly available within spreadsheet packages). This test is recommended in cases (which rarely arise in practice) where the standard deviation $\hat{\sigma}$ for the new tests is larger than the standard deviation σ for the main database.

1.7.4 *Method*

The null hypothesis is that the new results belong to the same population as the main database. The alternative hypothesis, that the new results belong to a population having longer fatigue lives than the main database, is accepted at the 5% level of significance if:

$$\overline{\log N_{\text{test}}} \geq \overline{\log N_D} + \frac{1.645\sigma}{\sqrt{n}} \tag{1.8}$$

where

$\overline{\log N_{\text{test}}}$ is the mean logarithm of the fatigue life from the tests at a particular stress.

$\overline{\log N_D}$ is the logarithm of the life from the mean S–N curve for the design class, and the value 1.645 is obtained from standard normal probability tables for a probability of 0.95.

Note that the corresponding level of significance of 5% is commonly considered to give a sufficiently low probability of concluding that the populations are different in the case where they are actually the same. For other significance levels, different values are obtained from the tables, e.g.

10% level of significance: 1.285
5% level of significance: 1.645
2.5% level of significance: 1.960
1% level of significance: 2.330.

Another way to express Eq. (1.8) is in the form:

$$g(N_{\text{test}}) \geq g(N_D) \times 10^{\left(\frac{1.645\sigma}{\sqrt{n}}\right)} \tag{1.9}$$

where $g(N)$ is the geometric mean of the appropriate fatigue lives.

Depending on the form of the information obtained from the fatigue tests, Eqs. (1.8) and (1.9) can be applied in a number of ways, some of which are detailed below.

1.7.5 Practical Applications

1.7.5.1 Tests Performed at the Same Stress Level

If all the fatigue tests are performed at the same stress level, $\overline{\log N_{\text{test}}}$ is the mean fatigue life obtained and n is the total number of tests. Unless the new results lie on an *S–N* curve with the same slope as the design curve, this approach only validates the class at the stress level used for the new tests.

1.7.5.2 Repeat Tests at Selected Stress Levels

If a number of tests are repeated at a few selected stress levels, Eq. (1.8) is applied in turn for each stress level, $\overline{\log N_{\text{test}}}$ being the mean life for each stress and n being the number of tests performed at the particular stress level considered. These tests validate the design curve over the selected range of stress levels, even if the *S–N* curve for the new test results does not have the same slope m as the main database.

1.7.5.3 Tests Performed to Produce a *S–N* Curve

If tests are performed at various stress levels and a *S–N* curve is fitted, Eq. (1.8) can be modified to compare this curve and the mean *S–N* curve for the design class being validated. A condition is that the curve fitted to the new results is assumed to have the same slope m as the design curve (see Sect. 1.7.3 (a)), giving an equation of the form $\log N + m \log S = .\ \overline{\log A_{\text{test}}}$ Then, Eq. (1.8) can be rewritten:

$$\overline{\log A_{\text{test}}} \geq \overline{\log A_D} + \frac{1.645\sigma}{\sqrt{n}} \tag{1.10}$$

or

$$g(A_{\text{test}}) \geq g(A_D) \times 10^{\left(\frac{1.645\sigma}{\sqrt{n}}\right)} \tag{1.11}$$

where n is the total number of tests. As a result, this approach is less demanding than that in Sect. 1.7.5.2 above because it relies on the *S–N* curves having the same slope.

As an example, consider the situation in which nine specimens are fatigue tested to failure to validate the use of Class D at the 5% level of significance:

Mean S–N curve for Class D: $S^3N = 3.99 \times 10^{12}$

Standard deviation of log N: $\sigma = 0.2097$

$n = 9$

Thus, the S–N curve fitted to the test results, assuming $m = 3$, must lie on or above the target S–N curve

$$S^3 N = A_{\text{target}}$$

where, from Eq. (1.11),

$$\begin{aligned} A_{\text{target}} &= 3.99 \times 10^{12}\ 10^{\left(\frac{1.645x0.2097}{3}\right)} \\ &= 5.2 \times 10^{12} \end{aligned}$$

In terms of the required endurances, this corresponds to achieving a mean S–N curve that is at least 1.3 times higher than the mean Class D S–N curve, or 3.42 times higher than the Class D design curve, which lies 2σ below the mean.

Note that the specimens fatigue tested do not necessarily need to fail for inequality (1.10) to be satisfied; they simply need to last longer (on average) than the lives obtained from the target S–N curve. This may be a more convenient approach than one in which the fatigue test conditions need to be chosen in order to establish an exact S–N curve. If, for instance, all the tests are designed to stop when a fixed value of $\log A_{\text{target}} = \log N + m \log S$ is reached, and none of the specimens fail, then the use of the design class can be justified using the target curve given by:

$$\log A_{\text{target}} \geq \overline{\log A_D} + \frac{1.645\sigma}{\sqrt{n}} \tag{1.12}$$

Note that, in this case, it would not be possible to test statistically the assumption of Sect. 1.7.3 (a) that the new results have the same slope m as the design curve.

Using the above example, but assuming that none of the nine specimens fails, the requirement would be that the endurance of each specimen must be at least 1.3 times higher than the corresponding mean Class D life.

1.8 Testing Whether Two Data Sets Are Consistent

1.8.1 Problem

It is often required to decide, on statistical grounds, whether two sets of S–N data can be regarded as forming part of the same population. For example, it may be necessary to test if a manufacturing change or the application of a post-weld fatigue life improvement technique really produces a significant change in fatigue performance. Similarly, the problem is likely to be of interest where the two data sets have been collected under conditions that are different (e.g. different research workers) but are expected to give comparable fatigue performance. The methods below can then be used to justify the amalgamation of the two data sets into one larger data set, or to justify the conclusion that there is a significant difference between them.

1.8.2 Approach

Initially, it is assumed that the two sets of test results follow the same *S–N* curve and have the same residual standard deviations about this curve (this so-called composite null hypothesis is, in effect, a combination of hypotheses). The observed differences are calculated between:

(a) the coefficients of the two regression lines through the two data sets, and
(b) the standard deviations $\hat{\sigma}_1$ and $\hat{\sigma}_2$ of the deviations (or "residuals") about these two regression lines.

Hypothesis testing is then used to check that, under these assumptions, it is not particularly unlikely (at a specified significance level) that the above statistics could have arisen by chance. This is the basis for regarding the null hypotheses as plausible, and for rejecting the alternative hypothesis that the two sets of test results belong to different populations.

Thus, the overall approach is similar to that in Sect. 1.7, the main difference being that here the "desired" outcome will probably be to confirm, rather than reject, the null hypotheses. For simplicity, it is assumed here that both sets of *S–N* data are exact data, although it is believed that the methods can, in principle, be extended to the case of censored data.

1.8.3 Tests Performed at the Same Stress Level

1.8.3.1 General

In this case, the coefficients of the "regression lines" referred to above simply reduce to the point estimates $\overline{\log N_1}$ and $\overline{\log N_2}$ of the mean logarithms of the fatigue lives. Also, the "residual standard deviations" reduce to simple standard deviations.

1.8.3.2 Test That Standard Deviations Are Consistent

The null hypothesis is that the two sets of results belong to populations having the same standard deviation. This hypothesis is accepted at a given significance level *a*% (e.g. the 5% level) if [19]:

$$\frac{\hat{\sigma}_1^2}{\hat{\sigma}_2^2} \leq F_{f_1}^{f_2} \tag{1.13}$$

where

$F_{f_1}^{f_2}$ is the $P\%$ percentage point (where $P = 100 - \alpha$) of the F-distribution (as obtained either from standard probability tables or software), which is a function of the numbers of degrees of freedom f_1 and f_2 used to estimate $\hat{\sigma}_1^2$ and $\hat{\sigma}_2^2$. As in Sect. 1.7, the numbers of degrees of freedom are one less than the corresponding sample sizes when the tests are all performed at the same stress level.

By convention, $\hat{\sigma}_1$ is taken to be the larger of the two standard deviations $\hat{\sigma}_1$ and $\hat{\sigma}_2$, while $\hat{\sigma}_2$ is taken to be the smaller of the two.

1.8.3.3 Test That the Mean Fatigue Lives Are Consistent

The null hypothesis is that the two sets of results belong to populations having the same mean fatigue life. This hypothesis is accepted at a given significance level *á*% (e.g. the 5% level) if [19]:

$$\left|\overline{\log N_1} - \overline{\log N_2}\right| \leq t\sqrt{\left(\frac{1}{n_1} + \frac{1}{n_2}\right)\sigma_e^2} \tag{1.14}$$

where

$\overline{\log N_1}$, $\overline{\log N_2}$ are the mean logarithms of the fatigue lives from the two sets of tests.
n_1 and n_2 are the sample sizes for the two sets of tests.
t is the two-sided $P\%$ percentage point (where $P = 100 - \alpha$) of Student's t distribution, with $(f_1 + f_2)$ degrees of freedom.
f_1 and f_2 are the numbers of degrees of freedom used to estimate $\hat{\sigma}_1^2$ and $\hat{\sigma}_2^2$, which are equal to $(n_1 - 1)$ and $(n_2 - 1)$, respectively, when the tests are all performed at the same stress level (as in Sect. 1.8.3.2).
σ_e^2 is an estimate of the common variance of the two samples, given by:

$$\sigma_e^2 = \frac{f_1\sigma_1^2 + f_2\sigma_2^2}{f_1 + f_2} \tag{1.15}$$

1.8.4 Tests Performed to Produce an S–N Curve

1.8.4.1 Test That Residual Standard Deviations Are Consistent

Equation (1.13) still applies in this case. But, as in Sect. 1.8.3.2, the numbers of degrees of freedom f_1 and f_2 (also used to estimate $\hat{\sigma}_1^2$ and $\hat{\sigma}_2^2$ are now equal to $(n_1 - 2)$ and $(n_2 - 2)$ respectively, because two coefficients have been estimated to obtain the *S–N* curves.

1.8.4.2 Test That the Intercepts of the Two *S–N* Curves Are Consistent

In this case, the equivalent formula to Eq. (1.14) is:

$$\left|\overline{\log A_1} - \overline{\log A_2}\right| \leq t\sqrt{\left(\frac{1}{n_1} + \frac{1}{n_2} + \frac{\left(\overline{\log S_1}\right)^2}{\sum_{i=1}^{n1}\left(\log S_{1,i} - \overline{\log S_1}\right)^2} + \frac{\left(\overline{\log S_2}\right)^2}{\sum_{j=1}^{n2}\left(\log S_{2,j} - \overline{\log S_2}\right)^2}\right)} \tag{1.16}$$

where

$\overline{\log A_1}$, $\overline{\log A_2}$ are the estimated intercepts of the regression lines through the two data sets.
n_1 and n_2 are the sample sizes for the two sets of tests.
t is the appropriate two-sided percentage point of Student's t distribution, with $(n_1 + n_2 - 4)$ degrees of freedom.
σ_e is an estimate of the common variance of the two samples given by Eq. (1.15).
$\overline{\log S_1}$ is the mean of the n_1 values of $\log S_i$ (i.e. the first data set).
$\overline{\log S_2}$ is the mean of the n_2 values of $\log S_j$ (i.e. the second data set).

1.8.4.3 Testing That the Slopes of the Two *S–N* Curves Are Consistent (*t*-Test)

In this case, the equivalent formula to equation [11, 26] is:

$$|m_1 - m_2| \leq t\sqrt{\left(\frac{1}{\sum_{i=1}^{n1}\left(\log S_{1,i} - \overline{\log S_1}\right)^2} + \frac{1}{\sum_{j=1}^{n2}\left(\log S_{2,j} - \overline{\log S_2}\right)^2}\right)\sigma_e^2} \tag{1.17}$$

where m_1 and m_2 are the estimated slopes of the regression lines through the two data sets, and all other notation is as in Sect. 1.8.4.2.

1.8.5 Composite Hypotheses

Note that if each of the three null hypotheses in Sects. 1.8.4.1–1.8.4.3 is tested at the 5% significance level, then there will be a probability of almost 15% that one of the three will be rejected even if all three are actually correct. For this reason, a lower (less demanding) significance level is often used for each individual test when a "composite" hypothesis such as this is tested, e.g. a significance level of 1.7% for each individual test would roughly correspond to a 5% significance level for the "composite" hypothesis. By a similar logic, it might be considered appropriate to choose a significance level of 2.5% for each of the two individual tests described in Sect. 1.8.3.

1.9 Testing Whether More than Two Data Sets Are Consistent

1.9.1 Problem

This section considers the extent to which the methods of Sect. 1.8 can be generalised to more than two sets of *S–N* data, i.e. the problem of deciding, on statistical grounds, whether the data sets can be regarded as forming part of the same population.

1.9.2 Approach

Initially the assumption is made that the M sets of test results follow the same *S–N* curve and have the same residual standard deviations about this curve (this combination of null hypotheses is another example of a "composite" hypothesis). The observed differences between the following are then assessed:

(a) the coefficients of the M regression lines through the two data sets, and
(b) the standard deviations $\hat{\sigma}_k$ $(k = 1, \ldots, M)$ of the residuals about the M regression lines.

Finally, hypothesis testing is used to check that, under these assumptions, it is not particularly unlikely (at a specified significance level) that the observed differences could have arisen by chance. This is the basis for regarding the null hypotheses as plausible, and for rejecting the alternative hypothesis that the M sets of test results belong to different populations.

Thus, the overall approach is analogous to that of Sect. 1.8. In particular, the observations of Sect. 1.8.5 concerning composite hypotheses also apply here. However, unless specifically stated otherwise, the methods of this section apply to exact data only.

1.9.3 Tests Performed at the Same Stress Level

1.9.3.1 General

In this case, the coefficients of the "regression lines" referred to above simply reduce to the point estimates $\overline{\log N_k}$ ($k = 1, \ldots, M$) of the mean logarithms of the fatigue lives, and the "residual standard deviations" reduce to simple standard deviations.

1.9.3.2 Test That Standard Deviations Are Consistent

The null hypothesis is that the M sets of results belong to populations having the same standard deviation. This hypothesis can be tested by direct application of Bartlett's test [7]. However, note that Bartlett's test is not robust to departures from normality. An alternative test, which is valid for any continuous distribution, is Levene's test, which is reported to be more robust for small samples [6].

1.9.3.3 Test That Mean Fatigue Lives Are Consistent

The null hypothesis is that the M sets of results belong to populations having the same mean logarithms for fatigue life. This hypothesis can be tested using a method known as single-factor Analysis of Variance (ANOVA) [19]. The ANOVA method is available within a wide range of software and (for exact data) from spreadsheet packages. It can also be performed (somewhat laboriously) by hand calculation, with reference to statistical tables of the percentage points of the F-distribution. For censored data, the ANOVA method can be applied as part of a "general linear model" (GLM) numerical procedure [6].

The ANOVA method identifies whether there are overall differences between the mean logarithms of the fatigue lives of the M data sets, but it does not, in itself, identify which data sets can be regarded as consistent and which ones are inconsistent with one another. Differences between pairs of data sets can be investigated using the t-test of Sect. 1.8.3.3. Also, there is a graphical analogue to the ANOVA method known as "Analysis of Means" (ANOM), which is particularly valuable for identifying if the mean logarithm of the fatigue life of one data set is significantly different from the other mean logarithms [6].

1.9.4 Tests Performed to Produce an S–N Curve

1.9.4.1 Test That Residual Standard Deviations Are Consistent

The same tests as in Sect. 1.9.3.2 apply here. However (as in Sect. 1.8.4.1), the number of degrees of freedom for each data set must be reduced by one accordingly. Note that this option was not available under the implementations of these tests in Release 12 of the MINITAB software package [6].

1.9.4.2 Test That the Intercepts and Slopes of the *S–N* Curves Are Consistent

The ANOVA method applies here, as in Sect. 1.9.3.3. Again, the main difference is that the numbers of degrees of freedom for each data set must be reduced accordingly. In this case, the use of hand calculations would be so laborious as to be rendered virtually impractical. However, GLM procedures are widely available that offer implementations of the ANOVA method and are also able to handle censored data. Differences between pairs of data sets can be investigated using the *t*-tests of Sect. 1.8.4.2 or 1.8.4.3, as appropriate. However, the authors are not aware of any implementations of the ANOM method in this particular case.

1.10 Sensitivity of Design Curve to Sample Size

For a given stress range *S*, the lower one-sided *P*% prediction limit, as given by Eq. (1.4), takes the general form:

$$\log N^{-}_{p\%} = \hat{\mu} - ts \tag{1.18}$$

where

$\hat{\mu}$ is an estimate of the mean endurance at stress *S*, based on *n* observations, *s* is an estimate of the standard deviation of the endurance at stress *S*, based on *f* degrees of freedom.
t is the appropriate percentage point of Student's *t* distribution with *f* degrees of freedom.
f is equal to *n* minus the number of estimated coefficients (as previously).

Both $\hat{\mu}$ and *s* are subject to sampling uncertainties (especially when the sample size is small), which, in turn, can affect the accuracy of Eq. (1.18). These sampling errors can be assessed by estimating a lower confidence limit of the form $\hat{\mu} - ks$ on the prediction limit $\log N^{-}_{p\%}$. This means a statement can be made of the form: "At least a proportion $P\%$ of the population is greater than $\hat{\mu} - ks$ with confidence $\gamma\%$". The statistic k is called a one-sided tolerance limit factor [18].

In the general case where the slope of the regression line is estimated from the data (i.e. $f = n - 2$), tolerance limit factors for the normal distribution are not readily available, either from standard statistical tables or from spreadsheet software. However, p. 117 of Owen [18] gives a formula for determining k, which requires the evaluation of both the $P\%$ percentage points of the normal distribution (which is readily available) and the $\gamma\%$ percentage points of the non-central t distribution. The 90, 95% and 99% percentage points of the non-central t distribution can, in turn, be evaluated using the formulae and associated tables on pp. 109–112 of Owen [18]. The required calculations are somewhat laborious and are outside the scope of this best practice guide. References [20, 21] contain further tables of the non-central t distribution. Owen [18] also tabulates k directly for sample sizes $n \leq 4$. Some statistical software packages also provide estimates of tolerance limits for the case $f = n - 2$ (for any given values of n, $\gamma\%$ and $P\%$) [6].

When a fixed value is assumed for the slope of the regression line estimated from the data, then the number of degrees of freedom $f = n - 1$. In this case, Owen [18] tabulates k directly, over a wide range of sample sizes, for $\gamma\% = 90$ and 95%, and $P\% = 90$, 95, 97.5, 99 and 99.9%. The cases likely to be of most interest in connection with fatigue analysis are $\gamma\% = 90\%$, with $P\% = 97.5\%$ or possibly $P\% = 95\%$; for convenience, the corresponding values of k are reproduced in Table 1.1.

Table 1.1 One-sided tolerance limit factors k for a normal distribution for $\gamma\% = 90\%$

Sample size, n	Value of k for $P\% =$ 95%	Value of k for $P\% =$ 97.5%
2	13.090	15.586
3	5.311	6.244
4	3.957	4.637
5	3.401	3.983
6	3.093	3.621
7	2.893	3.389
8	2.754	3.227
9	2.650	3.106
10	2.568	3.011
11	2.503	2.936
12	2.448	2.872
13	2.403	2.820
14	2.363	2.774
15	2.329	2.735
16	2.299	2.700

(continued)

Table 1.1 (continued)

Sample size, n	Value of k for $P\%=$ 95%	Value of k for $P\%=$ 97.5%
17	2.272	2.670
18	2.249	2.643
19	2.228	2.618
20	2.208	2.597
21	2.190	2.575
22	2.174	2.557
23	2.159	2.540
24	2.145	2.525
25	2.132	2.510
30	2.080	2.450
35	2.041	2.406
40	2.010	2.371
45	1.986	2.344
50	1.965	2.320
60	1.933	2.284
70	1.909	2.257
80	1.890	2.235
90	1.874	2.217
100	1.861	2.203
120	1.841	2.179
145	1.821	2.158
300	1.765	2.094
500	1.736	2.062
∞	1.645	1.960

Appendix: Statistical Analysis of Fatigue Data Obtained from Specimens Containing Many Welds

S. J. Maddox

Welded specimens used to obtain fatigue data invariably contain more than one potential site for fatigue crack initiation. For example, in a simple butt weld there are four weld toes from which fatigue cracks could propagate. Specimens used to investigate the fatigue performance of attachments normally include more than one. In view of this situation, the fatigue life obtained from a test on a specimen containing n nominally identical welds, all of which might fail, is the lowest of n possible fatigue lives. Thus, this life is less than the average life which would have been obtained if each individual weld had been tested to failure. Similarly, the mean *S–N* curve obtained from regression analysis of the fatigue test results obtained from several welded specimen:

$$S^m N = A \tag{1.19}$$

lies below that corresponding to failure in all the welds. Assuming that the fatigue lives are log-normally distributed (as is usually found to be the case for welded joints) an estimate of the average life from this higher *S–N* curve can be obtained using extreme value statistics.

Order statistics enable an estimate to be made of the expected smallest value of a random sample of n observations drawn from a normally distributed parent population. Tabulated values are available to represent the mathematical expression used [22, 23]. This Appendix is concerned only with the smallest overall value in a sample, which corresponds to the case $i = 1$. This actually relates to a parent normal distribution with zero mean and unit variance and gives the expected deviation of the smallest of n observations from the mean value of those n observations, expressed in numbers of standard deviations (standard deviation = $\sqrt{\text{variance}}$). An extract from the table is shown in Table 1.2; for a sample of, say, 5 (e.g. 5 identical welds per specimen), the expected deviation is – 1.163, so that the smallest value is 1.163 standard deviations below the mean. The standard deviation of the smallest of n randomly selected observations from a distribution is less than that of a single observation. However, again, statistical tables are available [24] to estimate it [1]; an extract is given in Table 1.3. Thus, for $n = 5$, the variance is 0.4475, whereas Table 1.2 relates to a variance of 1. Thus, the variance for the minimum of five samples is $0.4475 \times 1 = 0.4475$.

As an example, consider the fatigue test results obtained from tests on specimens incorporating three test welds. The minimum fatigue life from three observations is therefore known ($\log N_3$) together with the standard deviation of $\log N$ for those observations, σ_3. Thus, the statistical method is used in reverse to infer the value

Table 1.2 Extract from table of expected values of normal order statistics relevant to the lowest of a series of results ($i = 1$ in Ref. [22])

n	2	3	4	5	10	20	30
Expected deviation	– 0.564	– 0.846	– 1.029	– 1.163	– 1.539	– 1.867	– 2.04

Table 1.3 Extract from table of variances or order statistics relevant to the lowest of a series of results ($i = 1$ in Ref. [24])

n	2	3	4	5	10	20
Variance	0.6817	0.5595	0.4917	0.4475	0.3433	0.2757

corresponding to $n = 1$ (i.e. the mean life obtained from three times as many specimens each with a single weld, $\log N_1$) and the corresponding standard deviation of $\log N$, σ_1. Referring Tables 1.2 and 1.3. for $n = 3$, the expected deviation is 0.846 and $(\sigma_1)^2 = 0.5595$. Thus,

$$\log N_1 = \log N_3 + 0.846\sqrt{\frac{\sigma_3^2}{0.5595}} \tag{1.20}$$

and the corresponding standard deviation of $\log N$ is

$$\sqrt{\frac{\sigma_3^2}{0.5595}}$$

To illustrate the use of the technique, consider the fatigue test results obtained from 21 specimens each with three test welds given in Fig. 1.3. Regression analysis of all the data gave the equation:

$$S^{2.90}N = A \tag{1.21}$$

where $A = 2.02 \times 10^{12}$ as shown in Fig. 1.3, and the standard deviation of $\log N$ was 0.158.

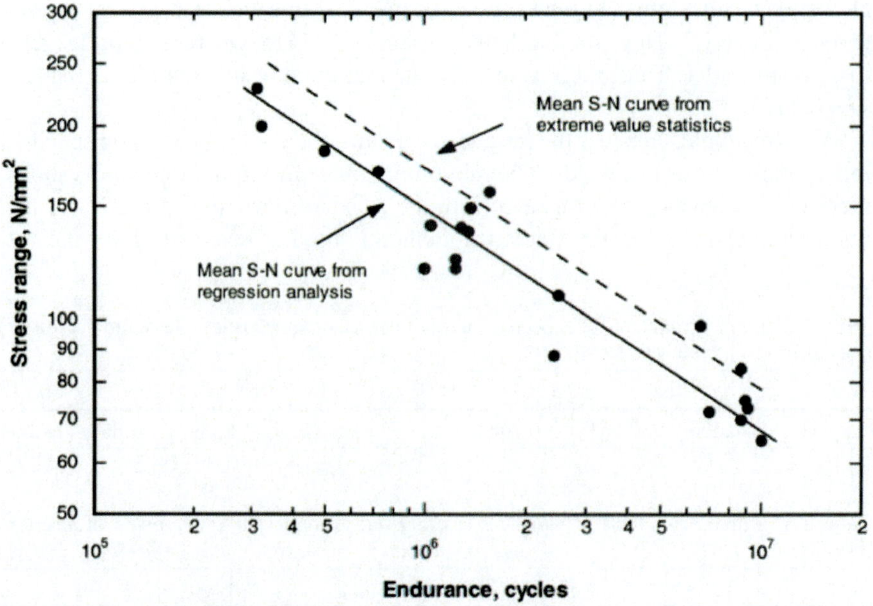

Fig. 1.3 Fatigue data obtained from specimens containing three welds, only one of which failed, analysed using extreme value statistics

Since N is proportional to A, Eq. (1.20) can be used directly to deduce the corresponding constant for the adjusted S–N curve for single welds:

$$\begin{aligned}\log A_1 &= \log A_3 + 0.846\sqrt{\frac{(0.158)^2}{0.5595}} \\ &= 12.306 + 0846\sqrt{\frac{0.0251}{0.5595}} \\ &= 12.845 \\ \text{or } A_1 &= 3.06 \times 10^{12}\end{aligned} \tag{1.22}$$

Thus, the equation of the adjusted mean S–N curve is $S^{2.90}\ N = 3.06 \times 10^{12}$ as shown in Fig. 1.3, representing a 52% increase in fatigue endurance at a stress range of 100 N/mm^2.

However, the standard deviation of log N has now increased to

$$\sqrt{\frac{0.0251}{0.5595}} = 0.212$$

Therefore, if an S–N curve some number of standard deviations below the mean was of interest the increase in fatigue endurance for single welds would be less. For example, for two standard deviations the increase is only 18%.

A further application of extreme value statistics is to deduce the average fatigue life of a structural member containing many welds any of which may fail. For example, for a member which incorporates 10 elements welded together in line, from Table 1.2 the **mean** fatigue life of such a member can be expected to be 1.539 standard deviations below the life obtained from any S–N curve deduced for single welds.

References

1. BS 2864: Guide to Statistical Interpretation of Data. BSI, London (1976)
2. ASTM: Practice for statistical analysis of linear or linearized stress-life (S–N) and strain-life (ε–N) fatigue data, ASTM E-739-91. Annual Book of ASTM Standards. ASTM, West Conshohocken, PA (1998)
3. BS 8118: Structural Use of Aluminium, Part 1 Code of Practice for Design. BSI, London (1991)
4. Hobbacher, A.: Fatigue Design of Welded Joints and Components. International Institute of Welding, Abington Publishing, Cambridge (1996), Section 3.7
5. Gurney, T.R., Maddox, S.J.: A re-analysis of fatigue data for welded joints in steel. Weld. Res. Int. **3**(4) (1973)
6. Minitab Reference Manual—Release 12 for Windows. Minitab Inc., USA (1998). ISBN: 0-925636-40-1
7. Cooper, B.E.: Statistics for Experimentalists. Pergamon, Oxford (1969). ISBN: 0 08 012600 6
8. Ferguson, G.A.: Statistical Analysis in Psychology and Education. McGraw-Hill, New York (1966), Table 21.4
9. Box, G.E.P., Tidwell, P.W.: Transformation of the independent variables. Technometrics **4**, 531–550 (1962)

10. D'Augostino, R.B., Stevens, M.A. (eds.): Goodness-of-Fit Techniques. Marcel Dekker (1986)
11. Shapiro, S.S., Wilk, M.B.: An analysis of variance test for normality (complete samples). Biometrika **52**, 591 (1965)
12. Maddox, S.J.: Fatigue assessment of welded structures. In: Proceedings of the International Conference on Extending the Life of Welded Structures, pp. 33–42. Pergamon, Press, Oxford, 1993
13. Maddox, S.J.: Fatigue design rules for welded structures. Prog. Struct. Eng. Mater. **2**(1), 102–109 (2000)
14. Bastenaire, F.: New method for the statistical evaluation of constant stress amplitude fatigue test results. In: Probabilistic Aspects of Fatigue, ASTM STP 511, pp. 3–28. ASTM, Philadelphia (1972)
15. Marquis, G., Mikkola, T.: Analysis of welded structures with failed and non-failed welds based on maximum likelihood. IIW document XIII-1822-2000, July 2000
16. Wallin, K.: The probability of success using deterministic reliability. In: Marquis, G., Solin, J. (eds.) Fatigue Design and Reliability. ESIS Publication, vol. 23, pp. 39–50. Elsevier Science Ltd, Amsterdam (1999)
17. Maddox, S.J.: Statistical analysis of fatigue data obtained from specimens containing many welds. IIW document XIII-XV-122-94 Rev 1, July 2006
18. Owen, D.B.: Handbook of Statistical Tables. Addison-Wesley, Reading, MA (1962)
19. Erricker, B.C.: Advanced General Statistics. ISBN: 0 340 15178 1
20. Johnson, N.L., Welch, B.L.: Applications of the non-central *t*-distribution. Biometrika **31**, 362–389 (1940)
21. Resnikoff, G.J., Lieberman, G.J.: Tables of the Non-central *t*-Distribution. Stanford University Press, Stanford, CA (1957)
22. Fisher, R.A., Yates, F.: Statistical Tables for Biological, Agricultural and Medical Research, 6th edn. Oliver & Boyd, Edinburgh (1963)
23. Beyer, W.H.: Handbook of Tables for Probability and Statistics, 2nd edn. CRC Press, FL (1986)
24. Pearson, E.S., Hartley, H.O.: Biometrika Tables for Statisticians. Cambridge University Press (1962)
25. BS EN 13445: Unfired Pressure Vessels, Part 3. BSI, London (2002)
26. Shapiro, S.S., Francia, R.S.: An approximate analysis of variance test for normality. J. Am. Stat. Assoc. **67**, 215 (1972)

Chapter 2
Working Sheets

2.1 Introduction

Background

Fatigue testing is widely used in research, product development and structural integrity assessment for welded structures. Many design recommendations for welded structures included fatigue testing as an important aspect of the fatigue design and assessment process.

The definition of the design fatigue strength of welded joints requires sample tests and statistical treatment of the results obtained. Fatigue testing is used to determine basic material or material joining characteristics such as:

- fatigue limit
- *S–N* fatigue strength curves
- crack propagation curves (m and C).

When developing a standard, the practice is to collate data from different sources to increase the number of results. In the past, this collation has usually been done without verification of the statistical equivalence of the collated sets of data.

If all the testing conditions are similar, fatigue test results present a dispersion which is considered as an intrinsic dispersion linked to material fatigue strength characteristics and has to be taken in account when determining design criteria.

A set of data that presents an intrinsic dispersion can be analysed and characterised using statistics. The parameters used are:

- mean and standard deviation (stdv)
- mean curve and stdv of the curve parameters
- curves at a specified probability level and confidence limits.

G. Parmentier et al., *Best Practice Guideline for Statistical Analyses of Fatigue Results*, IIW Collection, https://doi.org/10.1007/978-3-031-23570-2_2

Ideally, a confidence interval should be associated with the estimated mean and stdv, which is very rarely done in practice.

Objective

The present part provides worksheets to assist in the proper statistical assessment of experimental fatigue data as a complement of the above Part 1 “Best practice on statistical analysis of fatigue data”.

Each work sheet includes three parts: general, procedure and example of application.

2.2 Sheet 1: Can Two Data Sets Be Merged?

General

1. **Context**

To determine a *S–N* or d*a*/d*N* curve and increase the number of data to achieve greater precision in the statistical treatment, one solution is to collate results coming from two different test series performed on the same type of weld detail with same loading.

But such a procedure is acceptable only if the results correspond to the same global statistical family, i.e. if the differences between means and variances can be considered statistically negligible.

2. **Principle**

The verification of the statistical equivalence of the means or variances can be obtained using a hypothesis test.

Such test considers the hypothesis of the equivalence and provides a probability level that this hypothesis will be incorrectly rejected.

3. **Conditions of application**

The necessary data and conditions to perform the equivalence verification are the following:

- 2 data sets, $X1_i$, n_1 values and $X2_i$, n_2 values.
- All the data are "exact", i.e. there should be no censored data (such as run-outs).
- The data $X1_i$ and $X2_i$ have been verified following a normal distribution (see sheet 4 or 5). If the data sets do not follow a normal distribution, the number of data n_1 and n_2 shall be greater or equal to 30.
- The estimated means of the two data sets $X1_m$ and $X2_m$.
- The estimated variances of the two data sets v_1 and v_2.
- The variances have been verified to be equivalent (see sheet 2).

4. **Conclusion**

When the conditions of application are verified and the hypothesis of equivalence has no reason to be rejected, the two data sets can be merged to create a new data set of $n = n_1 + n_2$ values.

Procedure

1. **Criteria**

The applied test is the two-tailed Student's t-test for unpaired samples.

The selected level of risk is set to α, $\alpha\%$ probability to reject a correct hypothesis.

Note There is various definition of the function $t(\beta, \nu)$. The considered definition in the present sheet is so that

$$t(0.95, 10) = 1.812$$

2. **Data and formula**

	Sample No. 1	Sample No. 2
Sample size—n_i	……	……
Sum of values—ΣX_i	……	……
Mean value—$X_m = \frac{\sum X_i}{n_i}$	……	……
Sum of centred square values—$\Sigma (X_i - X_m)^2$	……	……
Estimated variances—$v_i = \frac{\sum (X_i - X_m)^2}{n_i - 1}$	……	……
Estimated common variance—$v = \frac{(n_1-1)v_1+(n_2-1)v_2}{(n_1+n_2-2)}$	……	
Estimated difference variance—$v_d = v\left(\frac{1}{n_1} + \frac{1}{n_2}\right)$	……	
Degree of freedom $\nu = (n_1 + n_2) - 2$	……	
$1 - \alpha/2$	…….	
Critical value for two-sided Student's t-test—$t\left(1 - \frac{\alpha}{2}, \nu\right)$	……	
Mean difference—$X_{m1} - X_{m2}$	……	
Criteria—$\mathrm{Cr} = \lvert X_{m1} - X_{m2} \rvert - t\left(1 - \frac{\alpha}{2}, \nu\right)\sqrt{v_d}$	……	

3. **Conclusion**

If $\mathrm{Cr} > 0$ the hypothesis of equivalence has to be rejected.

Example

1. **Data and formula**

Samples data

No. 1, $n_1 = 20$									
8	7.8	8	8	8	7.8	8	7.9	8	8.1
8	8.1	7.7	7.5	8	8	8.1	7.9	7.8	8.2
No. 2, $n_2 = 18$									
7.7	7.8	8.4	8.2	8	7.9	8.2	7.8	7.8	8.2
7.9	8.1	8	8.2	8.4	8.1	8.4	8.1		

2. **Calculations**

The risk level is fixed to $\alpha = 5\%$

	No. 1 sample	No. 2 sample
Sample size—n_i	20	18
Sum of values—ΣX_i	158.9	145.2
Mean value—$X_{mi} = \frac{\sum X_i}{n_i}$	7.945	8.067
Sum of centred square values—$\Sigma (X_i - X_m)^2$	0.4895	0.8200
Estimated variances—$v_i = \frac{\sum (X_i - X_m)^2}{n_i - 1}$	0.0258	0.0482
Estimated common variance—$v = \frac{(n_1-1)v_1+(n_2-1)v_2}{(n_1+n_2-2)}$	0.0364	
Estimated difference variance—$v_d = v\left(\frac{1}{n_1} + \frac{1}{n_2}\right)$	0.00384	
Degree of freedom $\nu = (n_1 + n_2) - 2$	36	
$1 - \alpha/2$	0.975	
Critical value for two-sided Student's *t*-test—$t\left(1 - \frac{\alpha}{2}, \nu\right)$	2.028	
Mean difference—$\lvert X_{m1} - X_{m2} \rvert$	0.122	
Criteria—$Cr = \lvert X_{m1} - X_{m2} \rvert - t\left(1 - \frac{\alpha}{2}, \nu\right)\sqrt{v_d}$	– 0.004	

3. **Conclusion**

As $Cr < 0$ the hypothesis of equivalence has no reason to be rejected, and therefore the data can be merged to form a new data set.

2.3 Sheet 2: Are the Variances of Two Data Sets Statistically Equivalent?

General

1. **Context**

To determine if two data sets can be merged it is necessary to verify that the variances of these data sets can be assumed equivalent.

Such equivalence can be also necessary when comparing the scatter of two types of fabrication techniques, the scatter being a good indicator of the quality variability.

2. **Principle**

The verification of the statistical equivalence of the variances can be obtained using a hypothesis test.

Such test considers the hypothesis of the equivalence and provides a probability level that this hypothesis will be incorrectly rejected.

3. **Conditions of application**

The necessary data and conditions to perform the equivalence verification are the following:

- 2 data sets, $X1_i$, n_1 values and $X2_i$, n_2 values.
- All the data are "exact", i.e. there should be no censored data (such as run-outs).
- The data $X1_i$ and $X2_i$ have been verified following a normal distribution (see sheet 4 or 5). If the data sets do not follow a normal distribution, the number of data n_1 and n_2 shall be greater or equal to 30.
- The estimated means of the two data sets $X1_m$ and $X2_m$.
- The estimated variances of the two data sets v_1 and v_2.

4. **Conclusion**

When the conditions of application are verified and the hypothesis of equivalence has no reason to be rejected, the two variances can be considered statistically equivalent.

Procedure

1. **Criteria**

The applied test is the two-sided Snedecor F-test with unpaired samples.

The selected level of risk is set to α, $\alpha\%$ probability to reject a correct hypothesis.

2. **Data and formula**

	Sample No. 1	Sample No. 2
Sample size—n_i	……	……
Sum of values—ΣX_i	……	……
Mean value—$X_m = \frac{\sum X_i}{n_i}$	……	……
Sum of centred square values—$\Sigma(X_i - X_m)^2$	……	……
Estimated variances—$v_i = \frac{\sum (X_i - X_m)^2}{n_i - 1}$	……	……
Degree of freedom $\nu_i = n_i - 1$	……	……
Variance ratio—$R = \frac{v_i}{v_j}$; v_i, the greater of v_1 and v_2	……	
Critical value for two-sided Snedecor F-test—$F\left(1 - \frac{\alpha}{2}, v_i, v_j\right)$	……	
Criteria—$\text{Cr} = R - F\left(1 - \frac{\alpha}{2}, v_i, v_j\right)$	……	

3. **Conclusion**

If $\text{Cr} > 0$ the hypothesis of equivalence has to be rejected.

Example

1. **Data and formula**

Samples data

No. 1, $n_1 = 20$									
8	7.8	8	8	8	7.8	8	7.9	8	8.1
8	8.1	7.7	7.5	8	8	8.1	7.9	7.8	8.2
No. 2, $n_2 = 18$									
7.7	7.8	8.4	8.2	8	7.9	8.2	7.8	7.8	8.2
7.9	8.1	8	8.2	8.4	8.1	8.4	8.1		

2. **Calculations**

The risk level is fixed to $\alpha = 5\%$

	No. 1 sample	No. 2 sample
Sample size—n_i	20	18
Sum of values—ΣX_i	158.9	145.2
Mean value—$X_{mi} = \frac{\sum X_i}{n_i}$	7.945	8.067
Sum of centred square values—$\Sigma(X_i - X_m)^2$	0.4895	0.8200
Estimated variances—$v_i = \frac{\sum (X_i - X_m)^2}{n_i - 1}$	0.02576	0.04824
Degree of freedom $\nu_i = n_i - 1$	19	17
Variance ratio—$R = \frac{v_1}{v_2}$ as $v_2 > v_1$	1.872	

(continued)

(continued)

	No. 1 sample	No. 2 sample
Critical value for two-sided Snedecor F-test—$F\left(1-\frac{\alpha}{2}, v_2, v_1\right)$	2.57	
Criteria—$\mathrm{Cr} = R - F\left(1-\frac{\alpha}{2}, v_2, v_1\right)$	− 0.70	

3. **Conclusion**

As $\mathrm{Cr} < 0$ the hypothesis of equivalence has no reason to be rejected, and therefore the two variances are statistically equivalent.

2.4 Sheet 3: Are the Means of Two Data Sets Statistically Equivalent?

General

1. **Context**

When comparing two different fabrication techniques or procedures, it is necessary to be sure that the mean strength differences obtained from testing are real and not only due to the sampling. So it is necessary to verify that the means of these data sets can be assumed statistically different.

2. **Principle**

The verification of the statistical equivalence or not of the means can be obtained using a hypothesis test.

Such test considers the hypothesis of the equivalence and provides a probability level that this hypothesis will be incorrectly rejected.

3. **Conditions of application**

The necessary data and conditions to perform the equivalence verification are the following:

- Two data sets, $X1_i$, n_1 values and $X2_i$, n_2 values.
- The number of data of one set at least has to be smaller than 20.
- All the data must be "exact", i.e. there should be no censored data (such as run-outs).
- Data $X1_i$ and $X2_i$ have been verified following a normal distribution (see sheet 4 or 5).
- The estimated means of the two data sets $X1_m$ and $X2_m$.
- The estimated variances of the two data sets v_1 and v_2.
- The variances have been verified to be statistically different or equivalent (see sheet 2)
  - if statistically different apply Procedure 1
  - if statistically equivalent apply Procedure 2.

4. **Conclusion**

When the conditions of application are verified and the hypothesis of equivalence has to be rejected, the two means can be considered statistically different.

Procedure 1—Variances Statistically Different

1. **Criteria**

The applied test is the Aspin–Welch test based on the two-sided Student law with unpaired samples.

The selected level of risk is set to α, $\alpha\%$ probability to reject a correct hypothesis.

Note There is various definition of the function $t(\beta, \nu)$. The considered definition in the present sheet is so that

$$t(0.95, 10) = 1.812$$

2. **Data and formula**

	Sample No. 1	Sample No. 2
Sample size—n_i	......	......
Sum of values—ΣX_i	......	......
Mean value—$X_m = \frac{\sum X_i}{n_i}$	......	......
Sum of centred square values—$\Sigma(X_i - X_m)^2$	......	......
Estimated variances—$v_i = \frac{\sum(X_i - X_m)^2}{n_i - 1}$	......	......
Verification that the variances are statistically different (sheet 2)		
Variance of the mean difference—$v_d = \left(\frac{v_1}{n_1} + \frac{v_2}{n_2}\right)$	......	
Intermediary calculation $\frac{1}{\upsilon} = \frac{v_1^2}{(n_1-1)(n_1 v_d)^2} + \frac{v_2^2}{(n_2-1)(n_2 v_d)^2}$	......	
Degree of freedom ν = nearest integer of υ	......	
Critical value for two-sided Student's t-test—$t\left(1 - \frac{\alpha}{2}, \upsilon\right)$	......	
Mean difference—$X_{m1} - X_{m2}$	......	
Criteria—$\mathrm{Cr} = \lvert X_{m1} - X_{m2}\rvert - t\left(1 - \frac{\alpha}{2}, \nu\right)\sqrt{v_d}$	......	

3. **Conclusion**

If $\mathrm{Cr} > 0$ the hypothesis of equivalence has to be rejected.

Example—Procedure 1

1. **Data and formula**

Samples data

No. 1, $n_1 = 16$									
487.5	519.8	485.2	503.1	505.1	500.1	470.3	499.4	504.9	494.7
473.3	480.3	500.2	499.7	480.7	488.3				
No. 2, $n_2 = 14$									
480.3	494.0	486.1	534.5	521.8	479.8	496.9	528.6	540.3	532.6
558.2	547.2	5554.0	486.1						

2. **Calculations**

The risk level is fixed to $\alpha = 5\%$	Sample No. 1	Sample No. 2
Sample size—n_i	16	14
Sum of values—ΣX_i	7892.6	7240.2
Mean value—$X_m = \frac{\sum X_i}{n_i}$	493.29	517.17
Sum of centred square values—$\Sigma(X_i - X_m)^2$	2627.1	10,786.8
Estimated variances—$v_i = \frac{\sum (X_i - X_m)^2}{n_i - 1}$	175.138	829.756

The variances have been verified to be statistically different with sheet 2

Cr	+ 1.81
Variance of the mean difference—$v_d = \left(\frac{v_1}{n_1} + \frac{v_2}{n_2}\right)$	70.2
Intermediary calculation $\frac{1}{\upsilon} = \frac{v_1^2}{(n_1-1)(n_1 v_d)^2} + \frac{v_2^2}{(n_2-1)(n_2 v_d)^2}$	1/17.7
Degree of freedom ν = nearest integer of υ	18
Critical value for two-sided Student's t-test—$t\left(1 - \frac{\alpha}{2}, \nu\right)$	2.101
Mean difference—$X_{m1} - X_{m2}$	– 23.87
Criteria—$\mathrm{Cr} = \lvert X_{m1} - X_{m2} \rvert - t\left(1 - \frac{\alpha}{2}, \nu\right)\sqrt{v_d}$	6.26

3. **Conclusion**

As $\mathrm{Cr} > 0$ the hypothesis of equivalence has to be rejected, the means can be considered statistically different.

Procedure 2—Variances Are Statistically Equivalent

1. **Criteria**

The applied test is the two-sided Student t-test with unpaired samples.

The selected level of risk is set to α, $\alpha\%$ probability to reject a correct hypothesis.

2. **Data and formula**

	Sample No. 1	Sample No. 2
Sample size—n_i	……	……
Sum of values—ΣX_i	……	……
Mean value—$X_m = \frac{\sum X_i}{n_i}$	……	……
Sum of squares—$\sum X_i^2$	……	……
Sum of centred square values—$\Sigma(X_i - X_m)^2$	……	……
Estimated variances—$v_i = \frac{\sum (X_i - X_m)^2}{n_i - 1}$	……	……

(continued)

(continued)

Verification that the variances are statistically equivalent (sheet 2)

Intermediary calculation $v = \frac{\sum(X_{i1}-X_{m1})^2+\sum(X_{i2}-X_{m2})^2}{n_1+n_2-2}$	……
Variance of the mean difference—$v_d = \left(\frac{1}{n_1}+\frac{1}{n_2}\right)$	……
Degree of freedom $\nu = n_1 + n_2 - 2$	……
Critical value for two-sided Student's t-test—$t\left(1-\frac{\alpha}{2}, v\right)$	……
Mean difference—$X_{m1} - X_{m2}$	……
Criteria—$\mathrm{Cr} = \lvert X_{m1} - X_{m2}\rvert - t\left(1-\frac{\alpha}{2}, \nu\right)\sqrt{v_d}$	……

3. **Conclusion**

If Cr > 0 the hypothesis of equivalence has to be rejected.

Example—Procedure 2

1. **Data and formula**

Samples data

No. 1, $n_1 = 20$									
8	7.8	8	8	8	7.8	8	7.9	8	8.1
8	8.1	7.7	7.5	8	8	8.1	7.9	7.8	8.2
No. 2, $n_2 = 18$									
7.7	7.8	8.4	8.2	8	7.9	8.2	7.8	7.8	8.2
7.9	8.1	8	8.2	8.4	8.1	8.4	8.1		

2. **Calculations**

The risk level is fixed to $\alpha = 5\%$	No. 1 sample	No. 2 sample
Sample size—n_i	20	18
Sum of values—ΣX_i	158.9	145.2
Mean value—$X_{mi} = \frac{\sum X_i}{n_i}$	7.945	8.067
Sum of centred square values—$\Sigma(X_i - X_m)^2$	0.4895	0.8200
Estimated variances—$v_i = \frac{\sum(X_i-X_m)^2}{n_i-1}$	0.02576	0.04824

The variances have been verified to be statistically equivalent (sheet 2)

Cr	– 0.7
Intermediary calculation $v = \frac{\sum(X_{i1}-X_{m1})^2+\sum(X_{i2}-X_{m2})^2}{n_1+n_2-2}$	0.0364
Variance of the mean difference—$v_d = \left(\frac{1}{n_1} + \frac{1}{n_2}\right)$	3.84×10^{-3}
Degree of freedom $\nu = n_1 + n_2 - 2$	36
Critical value for two-sided Student's *t*-test—$t\left(1 - \frac{\alpha}{2}, v\right)$	2.028
Mean difference—$X_{m1} - X_{m2}$	– 0.122
Criteria—$\mathrm{Cr} = \lvert X_{m1} - X_{m2}\rvert - t\left(1 - \frac{\alpha}{2}, \nu\right)\sqrt{v_d}$	– 0.0040

3. **Conclusion**

As $\mathrm{Cr} < 0$ the hypothesis of equivalence has no reason to be rejected and therefore the 2 means are statistically equivalent.

2.5 Sheet 4: Are the Data Normal Distributed Using Henry Graph?

General

1. **Context**

When analysing fatigue test results, it is necessary to verified that log(N) or log(K) are normally distributed.

The fitting is based on the Henry graph and the acceptability test is based on the normality Filliben's test.

2. **Principle**

The verification of the normal distribution of a parameter is performed using the Henry graph.

Such test is based on the verification of the linearity of the cumulative frequency distribution using a graph with adapted scales.

Such test considers the criteria function of the correlation factor and provides a probability level that this linearity hypothesis will be incorrectly rejected.

3. **Conditions of application**

The necessary data and conditions to determine the Henry graph are the following:

- the data set, n values X_i
- the X_i data classed in increasing values Y_i
- the mean value $\overline{Y}$ of Y_i distribution
- the calculation of the cumulative frequency distribution F_i
- the $\Phi^{-1}(F_i)$ values where Φ is the normal standard distribution (mean = 0, stdv = 1)
- the graph Y_i function of $\Phi^{-1}(F_i)$
- the Filliben's test.

4. **Conclusion**

The linearity is checked by looking at a linear regression of the obtained points.

When the straight line appears acceptable, may be taking out the extreme points, the normal distribution can be considered acceptable. Then it can be determined the mean and the standard deviation:

$$Y_i = a\Phi^{-1}(F_i) + b + k\varepsilon_i$$

Mean	$m = b$
Standard deviation	stdv $= a$

5. **Confidence test**

The verification of the non-acceptability of the normal distribution although it is true is done by the Filliben's test using the following correlation factor [1-3]:

$$R_P = \frac{\sum (Y_i - \overline{Y})\Phi^{-1}(F_i)}{\left[\sum (Y_i - \overline{Y})^2 \sum (\Phi^{-1}(F_i))^2\right]^{1/2}}$$

References

Vogel, R.M.: The probability plot correlation coefficient test for the normal, lognormal, and Gumbel distribution hypotheses. Water Resour. Res. **22**(4) (1986)

Ryan, Jr., T.A., Joiner, B.L.: Normal Probability Plots and Tests for Normality. Pennsylvania State University (1976)

Procedure

1. **Data and formula**

Test data, n values X_i and classification of the data in increasing value order Y_i.

Calculation of the F_i, of the mean $\overline{Y}$ of Y_i and of the $\Phi^{-1}(F_i)$.

Where Φ is the normal distribution, Y_i are the values of X_i classed in increasing order.

The selected level of risk α, $\alpha\%$ probability to reject a correct hypothesis.

Empirical Mean

The mean is obtained as follow:

- The data Y_i are ordered following the increasing values.
- The mean is so that:

$$\overline{Y} = \frac{\sum_{i=1}^{n} Y_i}{n}$$

Calculation Table

X_i	Y_i	$\sum n_i$	F_i	$\Phi^{-1}(F_i)$	$(Y_i - \overline{Y})\Phi^{-1}(F_i)$	$(Y_i - \overline{Y})^2$	$[\Phi^{-1}(F_i)]^2$
		1	$1/(n+1)$				
		2	$2/(n+1)$				
		i	$i/(n+1)$				
		n	$n/(n+1)$				
	$\overline{Y}$				Σ	Σ	Σ

Henry Graph

Drawing of the Henry graph [Y_i as a function of $\Phi^{-1}(F_i)$], determination of the straight line equation and of the correlation factor R^2 as illustrated below

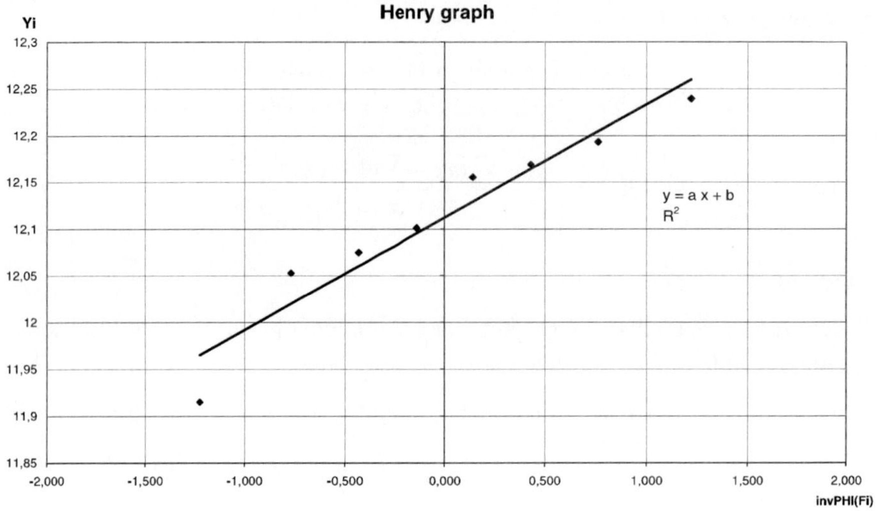

Note The calculations and graph drawing can be done with an Excel sheet

2. **Results**

Distribution mean value (Henry straight line constant), b	……
Distribution standard deviation (Henry straight line slope), a	……
$A = \sum (Y_i - \overline{Y})\Phi^{-1}(F_i)$	……
$B = \sum (Y_i - \overline{Y})^2$	……
$C = \sum [\Phi^{-1}(F_i)]^2$	……
$R_P = \frac{A}{\sqrt{BC}}$	……

Non-acceptance Criteria

The applied test is the Filliben's test.

The selected level of confidence is set to α, $\alpha\%$ probability to reject a correct assumption.

n		……
for $\alpha = 0.1$	$\text{Cv} = 1.0071 - \frac{0.1371}{\sqrt{n}} - \frac{0.3632}{n} + \frac{0.778}{n^2}$	……
for $\alpha = 0.05$	$\text{Cv} = 1.0063 - \frac{0.1288}{\sqrt{n}} - \frac{0.6118}{n} + \frac{1.3505}{n^2}$	……
for $\alpha = 0.01$	$\text{Cv} = 0.9963 - \frac{0.0211}{\sqrt{n}} - \frac{1.4106}{n} + \frac{3.1791}{n^2}$	……
$\text{Cr} = R_P - \text{Cv}$		……

2. **Conclusion**

if Cr < 0	the normality assumption is rejected
if Cr ≥ 0	the normality assumption is not rejected

If $\text{Cr} \geq 0$ the data can be assumed following a normal distribution with a probability α of not accepting a correct assumption.

Example

1. **Data and formula**

Fatigue test result data ($\log(K)$)

12.101 12.155 11.915 12.239 12.193 12.053 12.169 12.075

The selected level of risk α, probability to reject a correct hypothesis: 5%.

2. **Calculations**

Ordering $\log(K)$ values, calculation of the cumulative frequencies and of $\Phi^{-1}(F_i)$. The calculations have been done with an Excel sheet.

X_i	Y_i	$\sum n_i$	F_i	$\Phi^{-1}(F_i)$	$(Y_i - \overline{Y})\Phi^{-1}(F_i)$	$(Y_i - \overline{Y})^2$	$[\Phi^{-1}(F_i)]^2$
12.101	11.915	1	0.111	– 1.221	0.260	0.0454	1.4900
12.155	12.053	2	0.222	– 0.765	0.057	0.0056	0.5848
11.915	12.075	3	0.333	– 0.431	0.023	0.0028	0.1855
12.239	12.101	4	0.444	– 0.140	0.004	0.0007	0.0195
12.193	12.155	5	0.556	0.140	0.004	0.0007	0.0195
12.053	12.169	6	0.667	0.431	0.018	0.0017	0.1855
12.169	12.193	7	0.778	0.765	0.050	0.0042	0.5848
12.075	12.239	8	0.889	1.221	0.135	0.0123	1.4900
$\overline{Y}$	12.128				$\sum = 0.551$	$\sum = 0.0735$	$\sum = 4.5596$

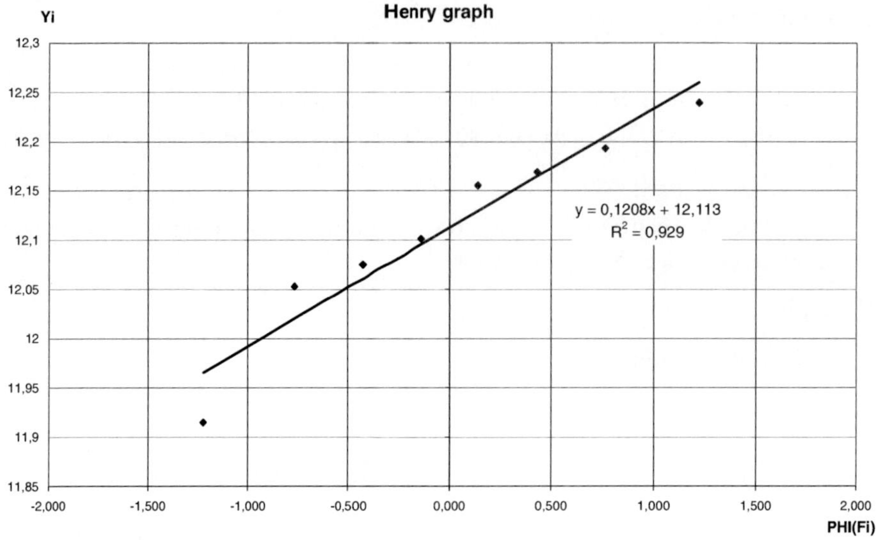

Distribution mean value (Henry straight line constant), b	12.113
Distribution standard deviation (Henry straight line slope), a	0.1208
$A = \sum (Y_i - \overline{Y})\Phi^{-1}(F_i)$	0.551
$B = \sum (Y_i - \overline{Y})^2$	0.0735
$C = \sum [\Phi^{-1}(F_i)]^2$	4.5596
$R_P = \frac{A}{\sqrt{BC}}$	0.951

Non-acceptance criteria

n	8
for $\alpha = 0.05$ $\text{Cv} = 1.0063 - \frac{0.1288}{\sqrt{n}} - \frac{0.6118}{n} + \frac{1.3505}{n^2}$	0.905
$\text{Cr} = R_P - \text{Cv}(0.05)$	0.046
Cr > 0, the data can be assumed following a normal distribution with $\alpha \leq 5\%$	

3. **Conclusion**

$\log(K)$ can be assumed normally distributed.

The mean value of $\log(K) = 12.113$.

The standard deviation of $\log(K) = 0.121$.

2.6 Sheet 5: Are the Data Normal Distributed Using a Likelihood Test?

General

1. **Context**

When analysing fatigue test results, it is necessary to verified that log(N) or log(K) are normally distributed.

The applied test is based on the fitting test of the data to a normal distribution.

2. **Principle**

The verification of the normal distribution of variable is performed using the K_i-two Pearson variable.

Such test is based on the fitting verification of the data to a normal distribution defined from the mean and standard deviation of the data set.

Such test provides a probability level that the fitting hypothesis will be incorrectly rejected.

3. **Conditions of application**

The necessary data and conditions are the following:

- the data set, N values X_i
- the mean and standard deviation of the data, the resulting normal distribution
- the determination of k classes to group the test results
- the number of elements resulting from the normal distribution in each class, no one including less than five elements.

4. **Conclusion**

When the conditions of application are verified and the criteria fulfilled, the hypothesis of normal distribution has no reason to be rejected.

Procedure

1. **Criteria**

The applied test is the K_i-two test.

The selected level of risk is set to α, $\alpha\%$ probability to reject a correct hypothesis.

2. **Data and formula**

Test data provide n_i values X_i in increasing order with a total of N values.

Sample size—N	……
Sum of values—$\Sigma n_i X_i$	……
Mean value—$X_m = \frac{\sum n_i X_i}{n_i}$	……

(continued)

(continued)

Estimated variances—$v = \frac{\sum n_i (X_i - X_m)^2}{N-1}$	……
Estimated standard deviation $s = v^{1/2}$	………

Theoretical values determination:

Initial data classes limits, a_i	$u_i = \frac{a_i - X_m}{s}$	$\Phi(u_i)$	$d_i = \Phi(u_i) - \Phi(u_{i-1})$	$t_i = Nd_i$	New classes, T_i	Test results, O_i	New classes
$-\infty$	–	0	–				
a_1			d_1	t_1			
a_2			d_2	t_2	$T_1 = t_1 + t_2$	$O_1 = n_1 + n_2$	1
a_3			d_3	t_3	$T_2 = t_3$	O_2	2
a_{N-1}			d_{N-2}	t_{N-2}	$T_{M-2} = t_{N-2}$	O_{M-2}	$M-2$
a_N			d_{N-1}	t_{N-1}	$T_{M-1} = t_{N-1} + t_N$	$O_{M-1} = n_{N-1} + n_N$	$M-1$
$+\infty$	–	1	d_N	t_N			

$a1 = \frac{X_1+X_2}{2}$ etc. $\Phi(x)$: cumulative normal law

Case where $t_1 < 5$ and $T_1 > 5$ $t_N < 5$ and $T_{M-1} > 5$

Degree of freedom $= M - 3$	…….
Critical value for K_i-two law—$\chi^2\left(1 - \frac{\alpha}{2}, v\right)$	……..
K_i-two law defined so that $\chi^2(1, 1) = 0$	
Distribution K_i-two variable—$\chi_D^2 = \sum \frac{(O_i - T_i)^2}{T_i}$	……
Criteria—$\mathrm{Cr} = \chi_D^2 - \chi^2\left(1 - \frac{\alpha}{2}, v\right)$	……

3. **Conclusion**

When $\mathrm{Cr} < 0$ the hypothesis of normal distribution has no reason to be rejected.

Example

1. **Data and formula**

Test result data

X_i	1	2	3	4	5	6
n_i	1	4	14	12	7	2

2. **Calculations**

The risk level is fixed to $\alpha = 5\%$

Sample size—$N = \Sigma n_i$	40
Sum of values—$\Sigma n_i X_i$	146
Mean value—$X_m = \frac{\Sigma n_i X_i}{N}$	3.65
Estimated variances—$v = \frac{\sum n_i (X_i - X_m)^2}{N-1}$	1.259
Estimated standard deviation $s = v^{1/2}$	1.122

Theoretical values determination:

Initial data classes limits a_i	$u_i = \frac{a_i - X_m}{s}$	$\Phi(u_i)$	$d_i = \Phi(u_i) - \Phi(u_{i-1})$	$t_i = N\, d_i$	New classes, T_i	Test results, O_i	New classes
$-\infty$	–	0	–				
1.5	– 1.916	0.0277	0.0277	1.107			
2.5	– 1.025	0.1527	0.1250	5.001	6.11	5	1
3.5	– 0.134	0.4468	0.2941	11.765	11.76	14	2
4.5	0.758	0.7756	0.3288	13.152	13.15	12	3
5.5	1.649	0.9504	0.1748	6.990	8.97	9	4
$+\infty$	–	1	0.0496	1.984			

$a1 = \frac{X_1 + X_2}{2}$ etc. … Φ(x): cumulative normal law.

Degree of freedom $\nu = M - 3$	1
Critical value for K_i-two law—$\chi^2(\alpha, \nu)$	3.84
K_i-two law defined so that $\chi^2(1, 1) = 0$	
Distribution K_i-two variable—$\chi_D^2 = \sum \frac{(O_i - T_i)^2}{T_i}$	0.727
Criteria—Cr $= \chi_D^2 - \chi^2(\alpha, v)$	– 3.115

3. **Conclusion**

As Cr < 0 the hypothesis of normal distribution has no reason to be rejected. The normal distribution of the data is accepted.

2.7 Sheet 6: Does a Standard *S–N* Curve Fit with a Data Set?

General

1. **Context**

When assessing a new design or new fabrication process, or a modification in fabrication process, it can be necessary to verified that the used design *S–N* curve remains valid.

The applied test is based on the validity of the hypothesis of identity of two distributions from two populations.

2. **Principle**

The verification of the equivalence of the test result data to the known *S–N* curve is performed using the Kruskal and Wallis H-test for 2 populations.

Such test is based on the analysis of the position of each data of each population in the total data set classed following increasing values.

3. **Conditions of application**

The necessary data and conditions are the following:

- the data sets, n_1 measures values X_i and n_2 values Y_i from the *S–N* curve
- a random number generation to determine the n_2 values of Y_i
- n_i must be greater than 5, although results remains acceptable with n_i until 3.

4. **Conclusion**

When the conditions of application are verified and the criteria fulfilled, the hypothesis of identity of the two distributions has no reason to be rejected. Therefore the *S–N* curve can be considered valid for the new measured data.

Procedure

1. **Criteria**

The applied test is the Kruskal and Wallis H-test.

The selected level of risk is set to α, $\alpha\%$ probability to reject a correct hypothesis.

2. **Data and formula**

Test data provide n_i values $X_i = \log(N_i)$ and the min ΔS_m and max ΔS_M values of ΔS_i.

The mean *S–N* curve is known: $\Delta S^m\ N = K$.

A number n_2 of values to be determined from the *S–N* curve is fixed, greater than 5 and n_1.

$(n_2 - 2)$ values u_i are determined from a random number generator or table, $u_i \in [0, 1]$ to which is added $u_1 = 0$ and $u_2 = 1$.

The values of $Z_i = \log(\Delta S_i)$ are calculated: $\Delta S_i = \Delta S_m + u_i\ (\Delta S_M - \Delta S_m)$.

The n_2 values $Y_i = \log(K) - m\log(\Delta S_i)$ are calculated. Excepting for $u_i = 0$ and 1, when $Y_i = a$ value X_j, this value is not considered and a new value of u_i is used.

Both X_i and Y_i are merged and classed in increasing values.

The position px_i of each value of X_i and py_i of Y_i in the merged series are determined and recorded as follows:

X_i/Y_i								
px_i								
py_i								

Total number of data $N = n_1 + n_2$	....
Sum $S_X = \Sigma px_i$	....
Sum $S_Y = \Sigma py_i$	......
Critical value for K_i-two law—$\chi^2(\alpha, 1)$	......
K_i-two law defined so that $\chi^2(1, 1) = 0$	
H parameter—$H = \frac{12}{N(N-1)}\left(\frac{S_X^2}{n_1} + \frac{S_Y^2}{n_2}\right) - 3(N+1)$	....
Criteria—$\text{Cr} = H - \chi^2(\alpha, 1)$	......

3. **Conclusion**

If $\text{Cr} < 0$ the hypothesis of identity of the distributions has no reason to be rejected. The S–N curve is acceptable for the new data.

Example

1. **Data and formula**

Fatigue test result data ($\log(K)$)

ΔS_i (MPa)	178	149	129	116	101
$X_i = \log(N_i)$	5.228	5.635	5.681	6.146	6.180

The design S–N curve is the UK DEn F2 curve: $m = 3$, $\log(K) = 12.092$.

2. **Calculations**

The risk level is fixed to $\alpha = 5\%$.

$n_1 = 5$, we shall take $n_2 = 6$.

Generation of the Y_i values:

$$\Delta S_m = 101 \quad \Delta S_M = 178 \qquad u_i \text{ from a table of random numbers}$$

u_i	0	1	0.849	0.503	0.180	0.385
ΔS_i	101	178	166.36	139.73	114.84	130.61
$Y_i = \log(N_i)$	6.079	5.341	5.429	5.656	5.912	5.744

X/*Y* classification and positions determination

X_i/Y_i	5.228	5.341	5.429	5.635	5.561	5.681	5.744	5.912	6.079	6.146	6.181
px_i	1			4		6				10	11
py_i		2	3		5		7	8	9		

Total number of data $N = n_1 + n_2$	11
Sum $S_X = \Sigma px_i$	32
Sum $S_Y = \Sigma py_i$	34
Critical value for K_i-two law—$\chi^2(\alpha, 1)$	3.84
K_i-two law defined so that $\chi^2(1, 1) = 0$	
H parameter—$H = \frac{12}{N(N-1)}\left(\frac{S_X^2}{n_1} + \frac{S_Y^2}{n_2}\right) - 3(N+1)$	0.133
Criteria—$\mathrm{Cr} = H - \chi^2(\alpha, 1)$	– 3.71

3. **Conclusion**

As $\mathrm{Cr} < 0$ the hypothesis of identity of the distributions has no reason to be rejected. The *S*–*N* curve is acceptable for the new data.

2.8 Sheet 7: Are Data Weibull Distributed Using a Weibull Probability Graph?

General

1. **Context**

The Weibull distribution is an important distribution especially for reliability and maintainability analysis. The graphical methods are used because of their simplicity and speed. However, it involves a great probability of error.

2. **Principle**

This test is based on the verification of the linearity of the cumulative frequency distribution using a graph with adapted scales.

The cumulative Weibull distribution function is given by:

$$F(x) = 1 - e^{-\left(\frac{x-\gamma}{\eta}\right)^{\beta}}$$

where β is the shape parameter, η is the scale parameter and γ is the location parameter.

With a change of variable it can be considered $\gamma = 0$ and the cumulative Weibull distribution function equation can be written as follows:

$$\ln\left[\ln\left(\frac{1}{1 - F(x)}\right)\right] = \beta(\ln x - \ln)$$

Anderson–Darling statistic is used to measure the non-parametric step function (based on the plot points). The Anderson–Darling AD value allows calculating an observed significance level OSL, probability p to be compared to the selected risk level α, $\alpha\%$ probability to reject a correct hypothesis.

If OSL < α, the hypothesis of Weibull distribution is rejected.

3. **Condition of application**

The necessary data and conditions to determine the Weibull probability graph are the following:

- data set, n values x_i (cycles) classed following increasing values
- $F(x_i)$. Commonly, it is chosen the median rank f to estimate the $F(x_i)$
- shape parameter k, scale parameter k
- AD value
- OSL level.

Procedure

1. **Criteria**

The selected level of risk is set to $k = 5\%$, probability to reject a correct hypothesis.

2. **Calculation**

For each x_i, $F(x_i)$ is estimated $F(x_i) = \frac{i-0.3}{n+0.4}$.

The shape parameter β and scale parameter η are calculated by using least square method:

Setting $Y = \ln\left[\ln\left(\frac{1}{1-F(x)}\right)\right]$		$X = \ln(x_i)$
$\ln\left[\ln\left(\frac{1}{1-F(x)}\right)\right] = \beta(\ln x - \ln)$	becomes	$Y = A + BX$

$\beta = B = \frac{n\sum X_i Y - \sum X_i \sum Y_i}{n\sum X_1^2 - (\sum X_i)^2}$	……
$A = \frac{\sum y_i}{n} - B\frac{\sum x_i}{n}$	……
$\eta = e^{-\left(\frac{A}{B}\right)}$	……

Calculation of the Anderson–Darling AD value:

Noting $Z_i = \left[\frac{x_i}{\eta}\right]^\beta$	
$\mathrm{AD} = \sum_{i=1}^n \frac{1-2i}{n}\{\ln[1-\exp(-Z_i)] - Z_{n+1-i}\} - n$	……
$\mathrm{AD}^* = \left(1 + \frac{0.2}{\sqrt{n}}\right)\mathrm{AD}$	……
Observed significant level OSL $= \frac{1}{1+\exp(-0.1+1.24\ln(\mathrm{AD}^*)+4.48\mathrm{AD}^*}$	……

3. **Conclusion**

If OSL > α the hypothesis of Weibull distribution has no reason to be rejected.

The Weibull distribution of the data is accepted.

Example

1. **Data and formula**

Long-term distribution $n = 28$

Cycles			
521,382	112,910	3,675,000	958,318
1,879,752	6,195,271	22,082,998	13,475,000
115,816	4,310,000	7,154,785	20,511,538
10,204,041	1,015,824	664,000	95,982
14,910,395	1,580,669	3,816,199	19,048,838

(continued)

(continued)

Cycles			
10,646,018	11,541,520	787,894	605,721
1,475,769	1,133,000	689,973	9,232,000

2. **Calculation**

Shape parameter β calculation by using least square method:

$$F(x_i) = \frac{i - 0.3}{n + 0.4}$$

$$Y = A + BX \qquad Y = \ln\left[\ln\left(\frac{1}{1 - F(x)}\right)\right] \qquad X = \ln(x)$$

x_i data	x_i ordered	i	$F(x_i)$	X	Y
521,392	95,982	1	0.0246	11.472	– 3.6906
1,879,752	112,910	2	0.0599	11.634	– 2.7851
115,816	115,816	3	0.0951	11.660	– 2.3036
10,204,041	521,392	4	0.1303	13.164	– 1.9691
14,910,395	605,721	5	.01655	13.314	– 1.7097
10,646,018	664,000	6	0.2007	13.406	– 1.4960
1,475,769	689,973	7	0.2359	13.444	– 1.3128
112,910	787,894	8	0.2711	13.577	– 1.1512
619,5271	958,318	9	0.3063	13.773	– 1.0057
4,310,000	1,015,824	10	0.3415	13.831	– 0.8726
1,015,824	1,133,000	11	0.3768	13.940	– 0.7490
1,580,669	1,475,769	12	0.4120	14.205	– 0.6330
11,541,250	1,580,669	13	0.4472	14.273	– 0.5230
1,133,000	1,879,752	14	0.4824	14.447	– 0.4177
3,675,000	3,675,000	15	0.5176	15.117	– 0.3161
22,082,998	3,816,199	16	0.5528	15.155	– 0.2172
7,154,785	4,310,000	17	0.5880	15.276	– 0.1201
664,000	6,195,271	18	0.6232	15.639	– 0.0241
3,816,199	7,154,785	19	0.6585	15.783	0.0716

(continued)

(continued)

x_i data	x_i ordered	i	$F(x_i)$	X	Y
787,894	9,232,000	20	0.6937	16.038	0.1681
689,973	10,204,041	21	0.7289	16.138	0.2663
958,318	10,646,018	22	0.7641	16.181	0.3676
13,475,000	11,541,250	23	0.7993	16.261	0.4737
20,511,538	13,475,000	24	0.8345	16.416	0.5871
95,982	14,910,395	25	0.8697	16.518	0.7120
19,048,938	19,048,938	26	0.9049	16.763	0.8557
605,721	20,511,538	27	0.9401	16.836	1.0352
9,323,000	22,082,998	28	0.9754	16.910	1.3092

$\beta = B = \frac{n\sum X_i Y - \sum X_i \sum Y_i}{n\sum X_1^2 - (\sum X_i)^2}$	0.7138
$A = \frac{\sum y_i}{n} - B\frac{\sum x_i}{n}$	– 11.034

A and *B* can be also obtained with Excel fitting a linear curve

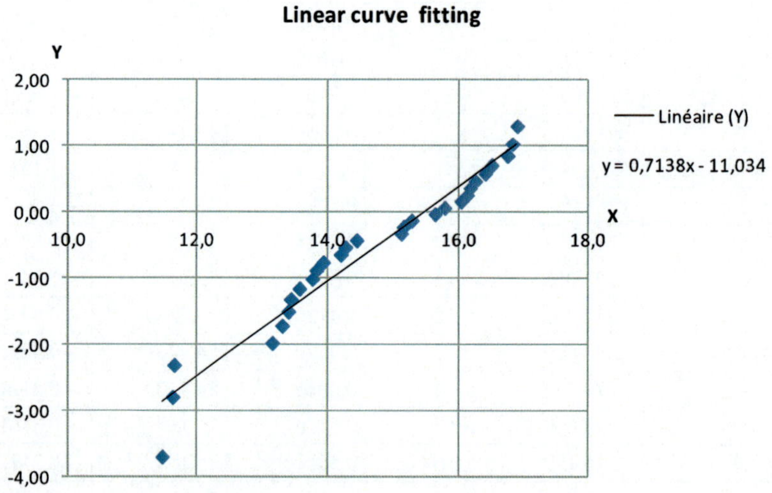

$\eta = e^{-\frac{A}{B}}$	5,166,858

Calculation of the Anderson–Darling AD value:

Noting $Z_i = \left[\frac{x_i}{\eta}\right]^{\beta}$	
$AD = \sum_{i=1}^{n} \frac{1-2i}{n}\left\{\ln\left[1-\exp(-Z_i)\right] - Z_{n+1-i}\right\} - n$	0.462
$AD^* = \left(1+\frac{0.2}{\sqrt{n}}\right)AD$	0.479
Observed significant level $OSL = \frac{1}{1+\exp\left[-0.1+1.24\ln(AD^*)+4.48AD^*\right]}$	0.243

3. **Conclusion**

As OSL > 0.05 the hypothesis of Weibull distribution has no reason to be rejected. The Weibull distribution of the data is accepted.

2.9 Sheet 8: How Many Results Are Necessary to Validate a Selected *S–N* Curve?

1. **Context**

In fatigue testing we would like to determine the number of samples to obtain reasonably precise estimates of the interested parameters but within a practical resource budget.

When having a set of n test results $(\Delta S_i, N_i)$, the objective is to verify if this number of data is sufficient to assure an acceptable level of precision for the estimated *S–N* curve.

2. **Principle**

An assumed *S–N* curve is selected: $\Delta S^m \, N = K$.

To determine the minimum sample size, the mean of the population is estimated.

When the sample data are collected and the sample mean and standard deviation are calculated, the margin of error δ, the maximum difference between the observed sample mean and the population mean is:

$$\delta = \delta_1 + \delta_2 = \overline{\log(K_{\text{test}})} - \log(K)$$

where

$\overline{\log(K_{\text{test}})}$ is the mean of $\log(K_i) = m \log(\Delta S_i) + \log(N_i)$.
$\log(K)$ is the $\log(K)$ of the assumed *S–N* curve.
$\delta_1 = \Phi(1-\alpha)\frac{\sigma}{\sqrt{n}}$ α: probability to reject a correct hypothesis.
$\delta_2 = \Phi(1-\beta)\frac{\sigma}{\sqrt{n}}$ β: probability to accept a wrong hypothesis.
n is the sample size.
σ is the $\log(K)$ standard deviation.
Φ is the standard normal distribution.

3. **Criteria**

The number of data (sample size) is acceptable when:

$$n \geq [\Phi(1-\alpha) + \Phi(1-\beta)^2 \left(\frac{\sigma}{\sqrt{n}}\right)^2 \qquad \text{One-sided test}$$

4. **Condition of application**

The necessary data and conditions are the following:

- The slope of the mean *S–N* curve for the test results is the same as m of the selected curve.
- The standard deviation of $\log(K_{\text{test}})$ is of same order than the standard deviation of $\log(K)$.

Procedure

1. Criteria

The corresponding values of $\alpha = 5\%$ and $\beta = 10\%$ are commonly considered to give a sufficiently low probability of concluding that the populations are different in the case where they are actually that same.

For other significance levels different values are obtained from the following table:

α/β (%)	$\Phi(1-\alpha)/\Phi(1-\beta)$
10	1.285
5	1.645
2.5	1.96
1	2.33

2. Data and formula

Test results: $n \quad (\Delta S_i, N_i)$.

Selection of a standard S–N curve: $\Delta S^m\, N = K$ standard deviation of $\log(K) = \sigma$.

Calculation of $\log(\Delta S_i)$, $\log(N_i)$, $\log(K_i)$.

Mean S–N curve determination by linear regression [$\log(\Delta S)$ function of $\log(N)$]

Slope of test curve	……
Slope of the standard S–N curve m	……

Calculation of $\log(K_i)$, mean $\log(K_i)$ and σ de $\log(K_i)$

Mean value of $\log(K_i)$, $\overline{\log(K_{\text{test}})}$	……
$\log(K)$	…….
Standard deviation of $\log(K_i)$	…….
Standard deviation of the standard S–N curve σ	…….
$\Phi(1-\alpha)$	…….
$\Phi(1-\beta)$	…….
$\delta = \overline{\log(K_{\text{test}}) - \log(K)}$	……

3. Criteria

The sample size is acceptable when:

$$n \geq [\Phi(1-\alpha) + \Phi(1-\beta)^2\left(\frac{\sigma}{\sqrt{n}}\right)^2$$

Example

1. **Data and formula**

The assumed *S–N* curve is Class D (UK DEn 1993):

mean curve $\Delta S^3\, N = K = 3.990 \times 10^{12}$ standard deviation of log(*K*) $\sigma = 0.2095$.

The considered rejection and acceptability levels are taken equal to:

$$\alpha = 5\% \quad \beta = 10\%$$

The obtained fatigue test results are the following: $n = 28$

Stress range	Cycles to failure	Stress range	Cycles to failure
147	521,382	74	3,675,000
96	1,879,752	53	22,082,998
250	115,816	53	7,154,785
61	10,204,041	136	664,000
57	14,910,395	75	3,816,199
57	10,646,018	136	787,894
96	1,475,769	136	689,973
250	112,910	147	958,318
74	6,195,271	53	13,475,000
74	4,310,000	53	20,511,538
136	1,015,824	265	95,982
136	1,580,669	54	19,048,838
53	11,541,520	176	605,721
136	1,133,000	74	9,232,000

2. **Calculation**

Calculation of $\log(\Delta Si)$, $\log(N_i)$, $\log(K_i)$.

Mean *S–N* curve determination by linear regression [$\log(\Delta S)$ function of $\log(N)$]

Slope of test curve	3.03
Slope of the standard *S–N* curve *m*	3

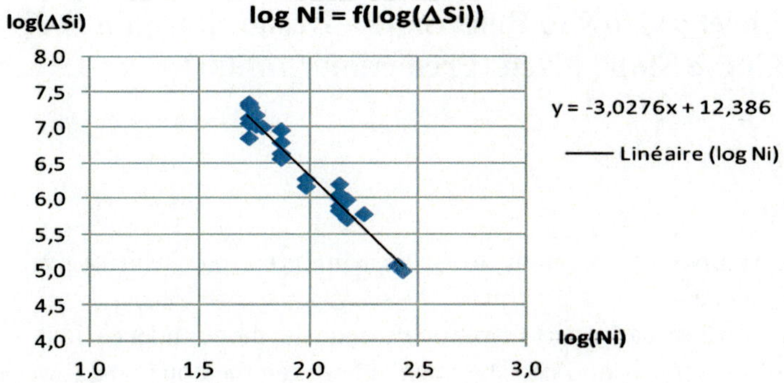

Calculation of $\log(K_i)$, mean $\log(K_i)$ and σ de $\log(K_i)$

Mean value of $\log(K_i)$, $\overline{\log(K_{\text{test}})}$	12.331
$\log(K)$	12.601
Standard deviation of $\log(K_i)$	0.1454
Standard deviation of the standard *S–N* curve σ	0.2095
$\Phi(1-\alpha)$	1.6649
$\Phi(1-\beta)$	1.2816
$\overline{\delta = \log(K_{\text{test}}) - \log(K)}$	− 0.269

3. **Criteria**

The minimum acceptable sample size is:

$n \geq [\Phi(1-\alpha) + \Phi(1-\beta)^2\left(\frac{\sigma}{\sqrt{n}}\right)^2$	5.2

4. **Conclusion**

According to above calculations the minimum number of sample size to validate the selected class D *S–N* curve should not be less than 6.

Therefore the fatigue test data are acceptable.

2.10 Sheet 9: How to Determine a Design Basquin *S–N* Curve Slope Fixed (Prediction Limits)?

General

1. **Context**

For design purpose it is necessary to establish limits between which a given proportion p of the data lies.

Prediction limits are used to avoid confusion with the confidence limits on the coefficients of the regression line. The interval between the upper and lower prediction limits is called a prediction interval.

2. **Principle**

Prediction limits at stress range S can be expressed explicitly in the form:

$$\log(N) = \left[\log(K) - m\log(S)\right] - t(p, f)\hat{\sigma}\sqrt{1 + \frac{1}{n}}$$

where

m, $\log(K)$ are the coefficients of the regression line through the n data points (S_i, N_i).
$t(p, f)$ is the centred Student's distribution value corresponding to a probability of p with f degrees of freedom.
$\hat{\sigma}$ is the best estimate of the standard deviation of the data about the regression line.
f is the degrees of freedom number.

3. **Condition of application**

- As slope m is chosen to take a fixed value, the number of degrees of freedom is equal to

$$f = n - 1$$

Procedure

1. **Criteria**

For welded details, in general, it can be used $m = 3$ and a proportion of non-failure $p = 0.975$ for one-sided Student law.

2. **Data and formula**

Test results: n data (N_i, S_i).

The mean curve estimated constant is expressed by: $\log(\widehat{K}) = \overline{\log(N_i)} + m\,\overline{\log(S_i)}$

Slope of the design curve fixed—m	……
Sample size—n	……
Sum of values—$\sum \log(S_i)$	……
Mean values—$\overline{\log(S)} = \frac{\sum \log(S_i)}{n}$	……
Sum of values—$\sum \log(N_i)$	……
Mean values—$\overline{\log(N)} = \frac{\sum \log(N_i)}{n}$	……
Estimated mean of regression line—$\log(\widehat{K})$	……
Calculate the estimated value—$\log(\widehat{N}_i) = \log(\widehat{K}) - m \log(S_i)$	
Degree of freedom $f = n - 1$	……
Estimated standard deviation—$\hat{\sigma} = \sqrt{\frac{\sum\left[\log(N_i)-\log\left(\widehat{N}_i\right)\right]^2}{f}}$	……
Centred Student's distribution corresponding to a probability of p with f degrees of freedom $t(p, f)$	……

Note If the used Student law is the two-sided one it must be calculated $t(2p - 1, f)$.

The function $t(p, f)$ table or function to be applied is not always clearly defined, therefore it must be verified that: $t(0.975, 10) = 2.228$. If not the value of p has to be adjusted.

Design *S–N* curve

$$\log(N) = \log(\widehat{K}) - t(p, f)\hat{\sigma}\sqrt{1 + \frac{1}{n}} - m \log(S)$$

Example

1. **Data and formula**

Samples data, $n = 9$

Stress range	Cycles to failure
147	521,382
96	1,879,752
250	115,816
61	10,204,041
57	14,910,395
57	10,646,018
96	1,475,769

(continued)

(continued)

Stress range	Cycles to failure
250	112,910
74	6,766,000

2. **Calculation**

Proportion of non-failure $p = 0.95$ one sided

The mean curve estimated constant is expressed by: $\log(\widehat{K}) = \overline{\log(N_i)} + m\,\overline{\log(S_i)}$

Slope of the design curve fixed—m	3
Sample size—n	9
Sum of values—$\sum \log(S_i)$	18.094
Mean values—$\overline{\log(S)} = \frac{\sum \log(S_i)}{n}$	2.010
Sum of values—$\sum \log(N_i)$	56.317
Mean values—$\overline{\log(N)} = \frac{\sum \log(N_i)}{n}$	6.257
Estimated mean of regression line—$\log(\widehat{K})$	12.2287
Calculate the estimated value—$\log(\widehat{N}_i) = \log(\widehat{K}) - m\log(S_i)$	
Degree of freedom $f = n - 1$	8
Estimated standard deviation—$\hat{\sigma} = \sqrt{\frac{\sum\left[\log(N_i) - \log\left(\widehat{N}_i\right)\right]^2}{f}}$	0.108
Centred Student's distribution corresponding to a probability of p with f degrees of freedom $t(p, f)$	2.306

Note If the used Student law is the two-sided one it must be calculated $t(2p - 1, f)$.

The function $t(p, f)$ table or function to be applied is not always clearly defined, therefore it must be verified that: $t(0.975, 10) = 2.228$. If not the value of p has to be adjusted.

Design S–N curve

$$\log(N) = \log(\widehat{K}) - t(p, f)\hat{\sigma}\sqrt{1 + \frac{1}{n}} - m\log(S)$$

$$\log(N) = 12.0264 - 3\log(S)$$

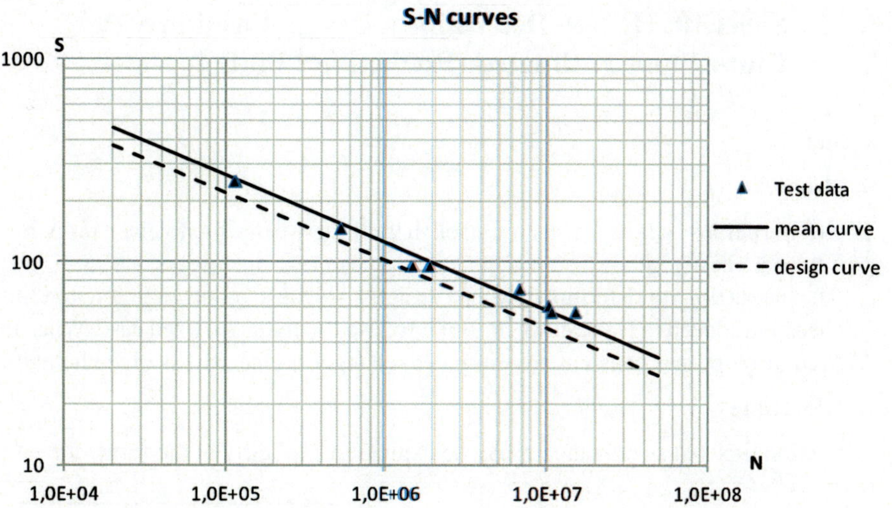
S-N curves
S
1000
100
10
N
1,0E+04
1,0E+05
1,0E+06
1,0E+07
1,0E+08
Test data
mean curve
design curve

2.11 Sheet 10: How to Determine a Design Basquin *S–N* Curve Slope Estimated (Prediction Limits)?

General

1. **Context**

For design purpose it is necessary to establish limits between which a given proportion p of the data lies.

The instruction for determination of a design *S–N* curve based on prediction limit has been introduced in Data Sheet 10. But it focuses on the slope fixed case. When the *S–N* curve slope and constant are estimated, changes are needed as given hereafter.

2. **Principle**

Prediction limits at stress range S can be expressed explicitly in the form:

$$\log(N) = \left[\log(K) - m\log(S)\right] - t(p,\, f)\hat{\sigma}\sqrt{1 + \frac{1}{n} + \frac{\left(\log(S) - \overline{\log(S_i)}\right)^2}{\sum\left(\log(S_i) - \overline{\log(S_i)}\right)^2}}$$

where

m and $\log(K)$ are the coefficients of the regression line through the n data points.
$\overline{\log(S_i)}$ is the mean of then n values of $\log(S_i)$.
log $t(p,f)$ is the centred Student's distribution value corresponding to a probability p with f degrees of freedom.
$\hat{\sigma}$ is the best estimate of the standard deviation of the data about the regression line.
f is the degree of freedom number.

The maximum likelihood estimation is used to estimate the *S–N* curve slope and constant.

3. **Condition of application**

- As slope m is estimated the number of degrees of freedom f is equal to $n - 2$.
- When the sample size larger than 20 the term in $\log(S)$ under the square root may be ignored.

Procedure

1. **Criteria**

For welded details, in general, it can be used a proportion of non-failure $p = 0.95$ for one-sided or $p = 0.9$ for two-sided interval.

2. **Data and formula**

The values for the constants m and K are obtained from the following two equations using the maximum likelihood method:

$\log(\widehat{K}) = \overline{\log(N_i)} + m\,\overline{\log(S_i)}$	$m = \frac{-\sum(\log(S_i)-\overline{\log(S_i)})(\log(N_i)-\overline{\log(N_i)})}{\sum(\log(S_i)-\overline{\log(S_i)})^2}$
Sample size—n	……
Sum of values—$\sum \log S_i$	……
Mean values—$\overline{\log(S_i)} = \frac{\sum \log(S_i)}{n}$	……
Sum of values—$\sum \log N_i$	……
Mean values—$\overline{\log(N_i)} = \frac{\sum \log(N_i)}{n}$	……
$\sum(\log(S_i) - \overline{\log(S_i)})(\log(N_i) - \overline{\log(N_i)})$	……
$\sum(\log(S_i) - \overline{\log(S_i)})^2$	……
Estimated slope—$\hat{m}$	……
Estimated mean of regression line $\log(\widehat{K})$	…….

Note The calculations and graph drawing can be done with Excel

Calculate the estimated value—$\log(\widehat{N}_i) = \log(\widehat{K}) - m\log(S_i)$

Degree of freedom $f = n - 2$	……
Estimated standard deviation—$\hat{\sigma} = \sqrt{\frac{\sum\left[\log(N_i)-\log\left(\widehat{N}_i\right)\right]^2}{f}}$	……
Centred Student's distribution corresponding to a probability of p with f degrees of freedom $t(p, f)$	……

Note If the used Student law is the two-sided one it must be calculated $t(2p - 1, f)$.

The function $t(p, f)$ table or function to be applied is not always clearly defined, therefore it must be verified that: $t(0.975,10) = 2.228$. If not the value of p has to be adjusted.

Design S–N curve

$$\log(N) = [\log(\widehat{K}) - m\log(S)] - t(p, f)\hat{\sigma}\sqrt{1 + \frac{1}{n} + \frac{(\log(S) - \overline{\log(S_i)})^2}{\sum(\log(S_i) - \overline{\log(S_i)})^2}}$$

The S–N curve is not linear in log-log; therefore in practice it is considered the S–N curve with $\log(S) = \overline{\log(S_i)}$

$$\log(N) = [\log(\widehat{K}) - m\log(S)] - t(p, f)\hat{\sigma}\sqrt{1 + \frac{1}{n}}$$

Example

1. **Data and formula**

Samples data, $n = 9$

Stress range	Cycles to failure
147	521,382
96	1,879,752
250	115,816
61	10,204,041
57	14,910,395
57	10,646,018
96	1,475,769
250	112,910
74	6,766,000

2. **Calculation**

Proportion of non-failure $p = 0.95$ one sided

Sample size—n	9
Sum of values—$\sum \log S_i$	18.041
Mean values—$\overline{\log(S_i)} = \frac{\sum \log(S_i)}{n}$	2.010
Sum of values—$\sum \log N_i$	59.317
Mean values—$\overline{\log(N_i)} = \frac{\sum \log(N_i)}{n}$	6.257
$\sum\left(\log(S_i) - \overline{\log(S_i)}\right)\left(\log(N_i) - \overline{\log(N_i)}\right)$	1.6913
$\sum\left(\log(S_i) - \overline{\log(S_i)}\right)^2$	0.5267
Estimated slope—$\hat{m} = \frac{-\sum\left(\log(S_i)-\overline{\log(S_i)}\right)\left(\log(N_i)-\overline{\log(N_i)}\right)}{\sum\left(\log(S_i)-\overline{\log(S_i)}\right)^2}$	3.21
Estimated mean of regression line $\log(\widehat{K}) = \overline{\log(N_i)} + m\overline{\log(S_i)}$	12.713

The calculations and graph drawing can be done with Excel:

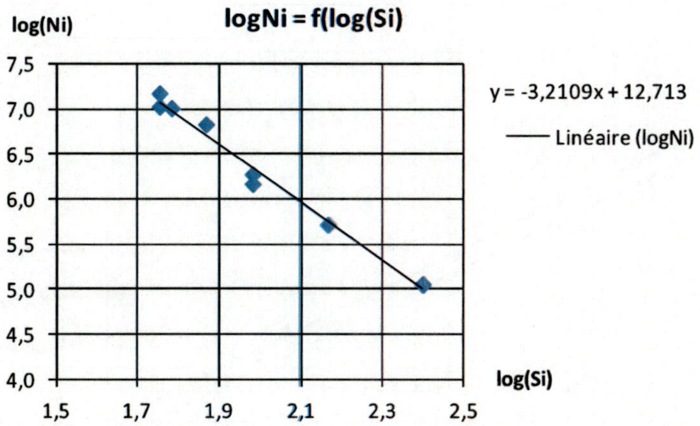

Calculate the estimated value—$\log(\widehat{N}_i) = \log(\widehat{K}) - m\log(S_i)$

Degree of freedom $f = n - 2$	7
Estimated standard deviation—$\hat{\sigma} = \sqrt{\frac{\sum\left[\log(N_i)-\log\left(\widehat{N}_i\right)\right]^2}{f}}$	0.10
Centred Student's distribution corresponding to a probability of p with f degrees of freedom $t(p, f)$	2.365

Note If the used Student law is the two-sided one it must be calculated $t(2p - 1, f)$.

The function $t(p, f)$ table or function to be applied is not always clearly defined, therefore it must be verified that: $t(0.975, 10) = 2.228$. If not the value of p has to be adjusted.

Design S–N curve

$$\log(N) = [\log(\widehat{K}) - m\log(S)] - t(p, f)\hat{\sigma}\sqrt{1 + \frac{1}{n} + \frac{\left(\log(S) - \overline{\log(S_i)}\right)^2}{\sum\left(\log(S_i) - \overline{\log(S_i)}\right)^2}}$$

$$\log(N) = 12.713 - 3.21\log(S) - 0.236\sqrt{1.111 + 1.8985[\log(S) - 2.010]^2}$$

In practice design S–N curve applied for design:

$$\log(N) = 12.713 - 3.21\log(S) - 0.2488 = 12.464 - 3.21\log(S)$$

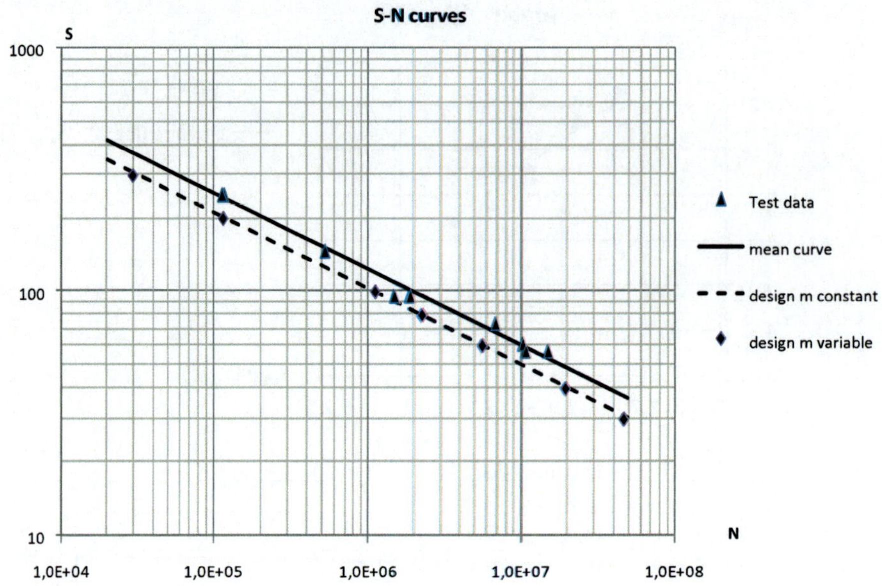
S-N curves
S
1000
100
10
1,0E+04
1,0E+05
1,0E+06
1,0E+07
1,0E+08
N
Test data
mean curve
design m constant
design m variable

2.12 Sheet 11: How to Determine a Design Basquin *S–N* Curve Slope Fixed (Tolerance Limits)?

General

1. **Context**

The use of tolerance limits rather than prediction limits would yield a more conservative design curve and they have the advantage that they explicitly allow for uncertainty in estimates of population statistics from a small sample.

2. **Principle**

This statement is made on the basis of a sample of n independent observations. A tolerance limit can be regarded as a confidence limit on a prediction limit:

$$\log(N) = \hat{\mu} - ks$$

where

$\hat{\mu}$ is an estimate of the mean $\log(N)$ at stress S.
s is an estimate of the standard deviation of the $\log(N)$ at stress S.
k is a one-sided tolerance limit factor based on f degrees of freedom.

The involved calculation to determine the one-side tolerance limit factor k is laborious. Mathematically, k is determined by the following equation:

$$k = \frac{t_{(f,\alpha)}(\delta)}{\sqrt{n}}$$

where

α is a confidence interval.
f is the number of freedom.
$t_{(f,\alpha)}(\delta)$ is the non-central Student law.
δ is defined by $\text{prob}\left\{\left(\text{noncentral } t \text{ with } \delta = \Phi^{-1}(p)\sqrt{n}\right) \leq k\sqrt{n}\right\} = \alpha$.
$\Phi^{-1}(p)$ is the inverse of the standard normal law: $\frac{1}{\sqrt{2\pi}} \int_{-\infty}^{\Phi^{-1}(P)} \exp\left(-\frac{x^2}{2}\right) \mathrm{d}x = p$.
p is a proportion of non-failure.

When f, α and δ are given, the critical value of the non-central Student distribution t can be calculated and then k.

Tolerance limits stated that at least a proportion of normal population is greater than $\hat{\mu} - ks$ with a confidence α.

3. **Condition of application**

The design curves are only to be applied to values of $\log(S)$ not far from the mean value $\overline{\log S}$.

Procedure

1. **Criteria**

For welded details, in general, it can be used a proportion of non-failure $p = 0.95$ for one-sided or $p = 0.90$ for two-sided interval and a confidence $\alpha = 0.75$ for bilateral interval or $\alpha = 0.875$ for unilateral interval.

2. **Data and formula**

Test results: n data (N_i, S_i)

The mean curve estimated constant is expressed by: $\log(\widehat{K}) = \overline{\log(N_i)} + m\,\overline{\log(S_i)}$

Slope of the design curve fixed—m	……
Sample size—n	……
Sum of values—$\sum \log(S_i)$	……
Mean values—$\overline{\log(S)} = \frac{\sum \log(S_i)}{n}$	……
Sum of values—$\sum \log(N_i)$	……
Mean values—$\overline{\log(N)} = \frac{\sum \log(N_i)}{n}$	……
Estimated mean of regression line—$\log(\widehat{K})$	……
Calculate the estimated value—$\log(\hat{N}_i) = \log(\widehat{K}) - m \log(S_i)$	
Degree of freedom $f = n - 1$	……
Estimated standard deviation—$s = \sqrt{\frac{\sum\left[\log(N_i) - \log\left(\widehat{N}_i\right)\right]^2}{f}}$	……
One-sided tolerance limit factor—$k(p, \alpha, n)$ (Appendix 1)	…….
Design curve constant $\log(K) = \log(\widehat{K}) - ks$	…….

Design S–N curve:

$$\log(N) = \log(K) - m \log(S)$$

Example

1. **Data and formula**

Samples data, $n = 9$

Stress range	Cycles to failure
147	521,382
96	1,879,752
250	115,816
61	10,204,041
57	14,910,395

(continued)

(continued)

Stress range	Cycles to failure
57	10,646,018
96	1,475,769
250	112,910
74	6,766,000

2. **Calculation**

Proportion of non-failure $p = 0.95$ one sided

Confidence $\alpha = 0.75$ bilateral

Slope of the design curve fixed—m	3
Sample size—n	9
Sum of values—$\sum \log(S_i)$	18.0941
Mean values—$\overline{\log(S)} = \frac{\sum \log(S_i)}{n}$	2.010
Sum of values—$\sum \log(N_i)$	53.3166
Mean values—$\overline{\log(N)} = \frac{\sum \log(N_i)}{n}$	6.257
Estimated mean of regression line—$\log(\widehat{K})$	12.2887
Calculate the estimated value—$\log(\widehat{N}_i) = \log(\widehat{K}) - m \log(S_i)$	
Degree of freedom $f = n - 1$	8
Estimated standard deviation—$s = \sqrt{\frac{\sum\left[\log(N_i) - \log\left(\widehat{N}_i\right)\right]^2}{f}}$	0.108
One-sided tolerance limit factor—$k(p, \alpha, f)$ (Appendix 1)	2.527849
Design curve constant $\log(K) = \log(\widehat{K}) - ks$	12.0156

Design S–N curve:

$$\log(N) = \log(K) - m \log(S)$$
$$\log(N) = 12.0156 - 3 \log(S)$$

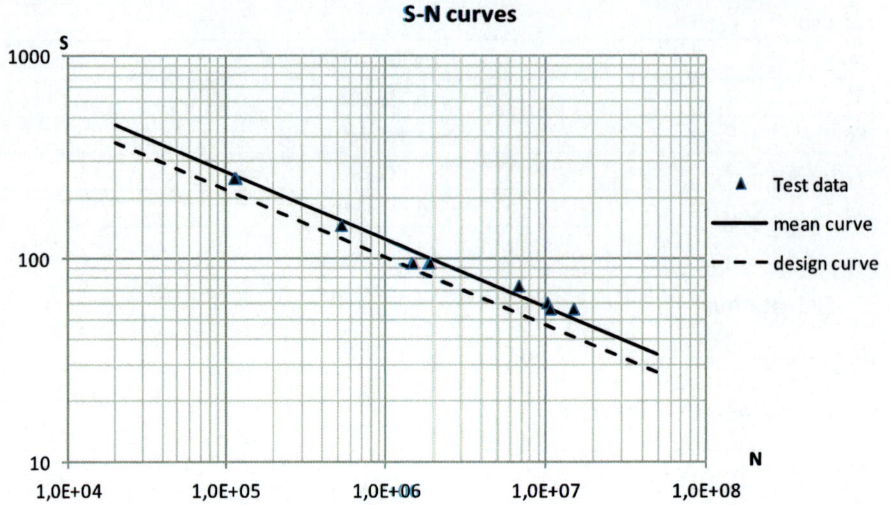
S-N curves
S
1000
100
10
1,0E+04
1,0E+05
1,0E+06
1,0E+07
1,0E+08
N
Test data
mean curve
design curve

2.13 Sheet 12: How to Determine a Design Basquin *S–N* Curve Slope Estimated (Tolerance Limits)?

General

1. **Context**

In Data Sheet 12 the instructions for determining a design *S–N* curve with fixed slope value based on tolerance limits have been illustrated. In the general case where the slope of the regression line is estimated from the data the tolerance limits factor is different from the previous example.

2. **Principle**

This statement is made on the basis of a sample of n independent observations. A tolerance limit can be regarded as a confidence limit on a prediction limit:

$$\log(N) = \hat{\mu} - ks$$

where

$\hat{\mu}$ is an estimate of the mean log(*N*) at stress *S*.
s is an estimate of the standard deviation of the log(*N*) at stress *S*.
k is an one-sided tolerance limit factor based on f degrees of freedom.

The involved calculation to determine the one-side tolerance limit factor k is laborious. Mathematically k is determined by the following equation:

$$k = \frac{t_{(f,\alpha)}(\delta)}{\sqrt{n^*}}$$

n^* is function of n and of the distance from the mean value of log(*S*).
α is a confidence interval.
f is the number of freedom.
$t_{(f,\alpha)}(\delta)$ is the non-central Student law.
δ is defined by $\text{prob}\left\{\left(\text{noncentral } t \text{ with } \delta = \Phi^{-1}(p)\sqrt{n^*}\right) \leq k\sqrt{n^*}\right\} = \alpha$.
$\Phi^{-1}(p)$ is the inverse of the standard normal law: $\frac{1}{\sqrt{2\pi}} \int_{-\infty}^{\Phi^{-1}(P)} \exp\left(-\frac{x^2}{2}\right)\mathrm{d}x = p$.
p is a proportion of non-failure.

When f, α and δ are given, the critical value of the non-central Student distribution t can be calculated and then k.

Tolerance limits stated that at least a proportion of normal population is greater than $\hat{\mu} - ks$ with a confidence α.

3. Condition of application

The design curves are only to be applied to values of $\log(S)$ not far from the mean value $\overline{\log S}$.

Procedure

1. Criteria

For welded details, in general, it can be used a proportion of non-failure $p = 0.95$ for one-sided or $p = 0.90$ for two-sided interval and a confidence $\alpha = 0.75$ for bilateral interval or $\alpha = 0.875$ for unilateral interval.

2. Data and formula

The values for the constants m and K are obtained from the following two equations using the maximum likelihood method:

$\log(\widehat{K}) = \overline{\log(N_i)} + m\overline{\log(S_i)}$	$m = \frac{-\sum\left(\log(S_i)-\overline{\log(S_i)}\right)\left(\log(N_i)-\overline{\log(N_i)}\right)}{\sum\left(\log(S_i)-\overline{\log(S_i)}\right)^2}$
Sample size—n	......
Sum of values—$\sum \log S_i$	......
Mean values—$\overline{\log(S_i)} = \frac{\sum \log(S_i)}{n}$	......
Sum of values—$\sum \log N_i$	......
Mean values—$\overline{\log(N_i)} = \frac{\sum \log(N_i)}{n}$	......
$\sum\left(\log(S_i) - \overline{\log(S_i)}\right)\left(\log(N_i) - \overline{\log(N_i)}\right)$	......
$\sum\left(\log(S_i) - \overline{\log(S_i)}\right)^2$	......
Estimated slope—$\hat{m}$	......
Estimated mean of regression line $\log(\widehat{K})$	....

Note The calculations and graph drawing can be done with Excel.
Calculate the estimated value—$\log(\widehat{N}_i) = \log(\widehat{K}) - m\log(S_i)$

Degree of freedom $f = n - 2$	......
Estimated standard deviation—$\widehat{\sigma} = \sqrt{\frac{\sum\left[\log(N_i)-\log\left(\widehat{N}_i\right)\right]^2}{f}}$	......

Design S–N curve:

Select a series of S

$$\text{Calculate}\,\frac{n^*}{n}(S) = \frac{\sum\left(\log(S_i) - \overline{\log(S_i)}\right)^2}{n\left(\log(S) - \overline{\log(S_i)}\right)^2 + \sum\left(\log(S_i) - \overline{\log(S_i)}\right)^2}$$

Determine the one-sided tolerance limit factor $k(p, \alpha, n, \frac{n^*}{n})$ (Appendix 2)

$$\log(N) = \log(\widehat{K}) - \hat{m}\log(S) - k(p, \alpha, n, \frac{n^*}{n})\widehat{\sigma}$$

The S–N curve is not linear in log-log, therefore in practice it is considered the S–N curve with $\frac{n^*}{n} = 1$

$$\log(N) = \log(\widehat{K}) - \hat{m}\log(S) - k(p, \alpha, n, 1)\widehat{\sigma}$$

Example

1. **Data and formula**

Samples data, $n = 9$

Stress range	Cycles to failure
147	521,382
96	1,879,752
250	115,816
61	10,204,041
57	14,910,395
57	10,646,018
96	1,475,769
250	112,910
74	6,766,000

2. **Calculation**

Proportion of non-failure $p = 0.95$ one sided

Confidence $\alpha = 0.75$ bilateral.

The values for the constants m and K are obtained from the following two equations using the maximum likelihood method:

$\log(\widehat{K}) = \overline{\log(N_i)} + m\overline{\log(S_i)}$	$m = \frac{-\sum\left(\log(S_i) - \overline{\log(S_i)}\right)\left(\log(N_i) - \overline{\log(N_i)}\right)}{\sum\left(\log(S_i) - \overline{\log(S_i)}\right)^2}$
Sample size—n	9
Sum of values—$\sum \log S_i$	18.0941
Mean values—$\overline{\log(S_i)} = \frac{\sum \log(S_i)}{n}$	2.010
Sum of values—$\sum \log N_i$	56.3166
Mean values—$\overline{\log(N_i)} = \frac{\sum \log(N_i)}{n}$	6.257

(continued)

(continued)

$\sum\left(\log(S_i) - \overline{\log(S_i)}\right)\left(\log(N_i) - \overline{\log(N_i)}\right)$	1.6913
$\sum\left(\log(S_i) - \overline{\log(S_i)}\right)^2$	0.5267
Estimated slope—$\hat{m}$	3.21
Estimated mean of regression line $\log(\widehat{K})$	12.713

The calculations and graph drawing can be done with Excel:

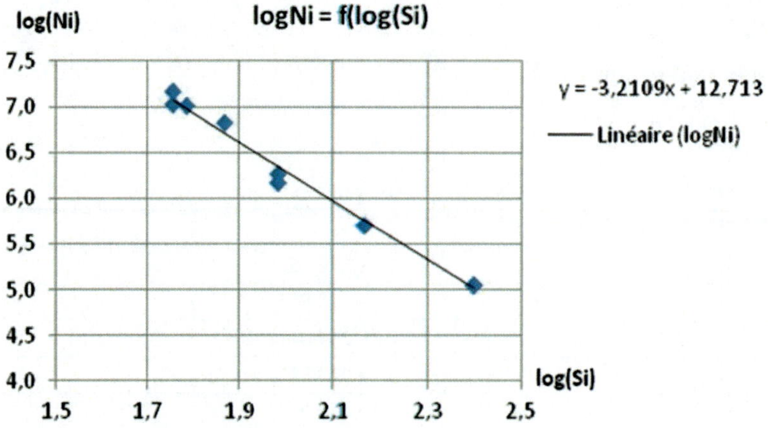

Calculate the estimated value—$\log(\widehat{N}_i) = \log(\widehat{K}) - m\log(S_i)$

Degree of freedom $f = n - 2$	7
Estimated standard deviation—$\hat{\sigma} = \sqrt{\frac{\sum\left[\log(N_i) - \log\left(\widehat{N}_i\right)\right]^2}{f}}$	0.10

Design S–N curve:

Select a series of S

$$\text{Calculate}\,\frac{n^*}{n}(S) = \frac{\sum\left(\log(S_i) - \overline{\log(S_i)}\right)^2}{n\left(\log(S) - \overline{\log(S_i)}\right)^2 + \sum\left(\log(S_i) - \overline{\log(S_i)}\right)^2}$$

Determine the one-sided tolerance limit factor $k(p, \alpha, n, \frac{n^*}{n})$ (Appendix 2) ($p = 0.95$ one sided, $\alpha = 0.75$ bilateral)

$$\log(N) = \log(\widehat{K}) - \hat{m}\log(S) - k(p, \alpha, n, \frac{n^*}{n})\hat{\sigma}$$

S_i	log S_i	n^*/n	k	log N
300	2.477	0.2118	2.972	4.464025495
250	2.398	0.2805	2.873	4.728096412
200	2.301	0.4094	2.761	5.050376555
150	2.176	0.6808	2.656	4.461928963
100	2.000	0.9981	2.602	6.032581423
50	1.699	0.3763	2.785	6.980589341
30	1.477	0.1706	3.088	7.662426529

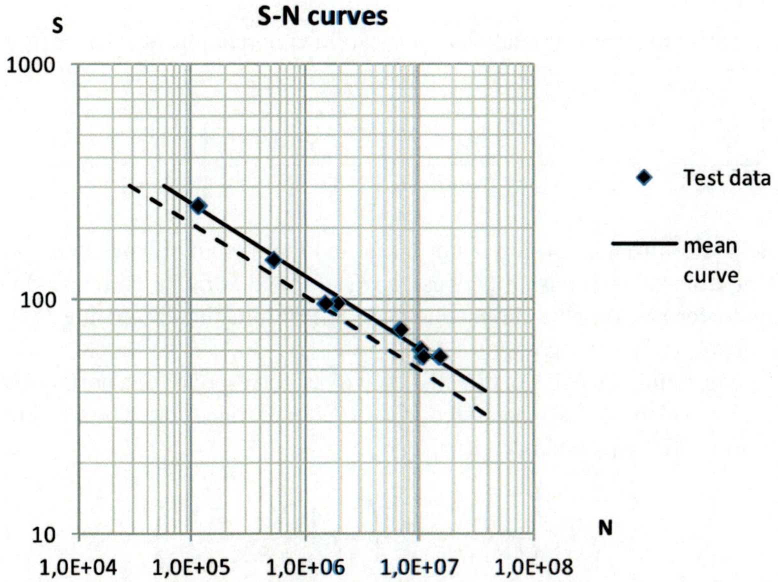

In practice design S–N curve applied for design:

$$\log(N) = \log(\widehat{K}) - \hat{m}\log(S) - k(p, \alpha, n, 1)\widehat{\sigma}$$

$$\log(N) = 12.7128 - 3.21\log(S) - 2.602x0.1 = 12.453 - 3.21\log(S)$$

2.14 Sheet 13: How to Determine a Mean *S–N* Curve Bastenaire Equation?

General

1. **Context**

In Data Sheets 9–12 the *S–N* curves are determined only with failed samples, so no information is obtained about the low cycle domain or about the fatigue limit.

The present method allows taking into account non-failed samples and provides an equation including the low cycle domain and an estimation of the fatigue limit.

2. **Principle**

As alternative to the linear relationship, the following relationship is used (S = stress range):

$$N = \frac{A}{S - E} \exp\left[-\left(\frac{S - E}{B}\right)^C\right]$$

The Bastenaire relationship is not linear and there is no obvious transformation allowing a linear relationship between a transformed variable corresponding to N and a transformed variable corresponding to S. It results that estimating A, B, C and E requires a nonlinear regression.

The uncertainty on N is not directly known but corresponds to a normal Gaussian model on S which can be expressed by $S = \Phi(N)$ and transformed by the Bastenaire relationship. The equation becomes:

$$N = \frac{A}{\Phi(N) - E} \exp\left(-\left(\frac{\Phi(N) - E}{B}\right)^C\right)$$

with $\Phi(N) \geq E \quad A \geq 0 \quad B \geq 0 \quad C \geq 0 \quad E \geq 0$.

The function $\Phi(N)$ is so an implicit function corresponding to the minimum distance between ΔS_i, stress range at level i, and $\Phi(N_i)$.

The A, B, C and E values correspond to:

$$\min \sum_{i=1}^{n} [S_i - \Phi(N_i)]^2$$

where N_{mi} is the median of the number of cycles corresponding to the stress range level S_i.

3. **Condition of application**

At least four stress range levels are necessary to estimate the four curve parameters.

To obtain a satisfactory result enough data are necessary on each stress range level. A number of 10–20 data are recommended.

Procedure

1. **Method**

The *S–N* curve is taken as the best fitting curve which minimises the distances between the curve and the median values of the number of cycles *N* at each stress range level *S*. The determination of *A*, *B*, *C*, *E* is very delicate and requires a specific software. Two formulations are given in this sheet:

- The first for a particular case which can be applied with Excel.
- The second applying the software R which can be downloaded from the web.

2. **Data**

The fatigue tests are performed at constant stress range S_i until failure or stopped when N_i reaches a fixed value, in principle 2×10^6 cycles.

The data are classed (S_i, N_{ij}) noting (nf) the non-failed samples.

i		1	2	3	4	*i*
N_{ji}	$j \backslash S_i$	S_1	S_2	S_3	S_4	S_i
	1	N_{11}	N_{12}			
	2	N_{21}				
	j	N_{j1}				N_{ji}

3. **Using Excel**

Excel can be used when the number of stress range level equal four as it is possible to have a system of four equations to determine the four variables *A*, *B*, *C*, *E*.

Mean N Values

The first step is to estimate the mean value of N_{mi} for each ΔS_i:

$$\log(N_{mi}) = \frac{1}{n\,i} \sum_{j=1}^{n} \log(N_{ji})$$

where n_i is the number of data items at level ΔS_i.

If the S_i level presents non-crack samples, the mean has to be determined with the log-log probability graph:

- for each ΔS_i order N_{ji} the in increasing levels
- calculated the probabilities p_{ji} for each N_{ji}: $p_{ji} = \frac{j-03}{n_i+0.4}$ or $p_{ji} = \frac{j}{n_i+1}$

  Note the two formula generates almost identical results when *n* is sufficiently large.

- for i, plot $\log(N_{ji})$ versus $\Phi^{-1}(p_{ji})$ and determine the best fitting line
- calculate $\log(N_{mi})$ corresponding to $\Phi^{-1}(0.5) = 0$ and N_{mi}.

Mean S–N Curve

The estimation of the mean curve is performed by iteration considering a modification of the Bastenaire formula as follows:

$$\frac{1}{N} = \frac{S-E}{A}\exp\left(\left(\frac{S-E}{B}\right)^C\right) = \Phi(S)$$

and the development of Φ as follows:

$$\Phi(S) = \Phi'_A\delta A + \Phi'_B\delta B + \Phi'_C\delta C + \Phi'_E\delta E$$

The iteration steps are as follows:

1. Select a set of values A_0, B_0, C_0 and E_0. The following values can be generally chosen:

$$A_0 = 1.0\,E06 \quad B_0 = 500 \quad C_0 = 1.0$$

$E_0 = S_{\min} - 1$ (MPa) $S_{\min}$ is the lower level with at least one failed specimen.
Verify by graph that the Bastenaire curve is below the test data points, if not adjust the values A_0, B_0, C_0 to obtain a curve below the data points.

2. For each S_i level

At the step k of the iteration (k starting $= 1$):

$$A = A_{k-1} \quad B = B_{k-1} \quad C = C_{k-1} \quad E = E_{k-1}$$

following functions are calculated:

$$\Phi(S_i) = \frac{S-E}{A}F(S_i) \quad \text{with } F(S_i) = \exp\left[\left(\frac{S_i-E}{B}\right)^C\right]$$

$$\Phi'_A(S_i) = -\frac{S_i-E}{A^2}F(S_i)$$

$$\Phi'_B(S_i) = -\frac{C}{A}\left(\frac{S_i-E}{B}\right)^{C+1}F(S_i)$$

$$\Phi'_C(S_i) = \frac{B}{A}\left(\frac{S_i-E}{B}\right)^{C+1}\ln\left(\frac{S_i-E}{B}\right)F(S_i)$$

$$\Phi'_E(S_i) = -\frac{1}{A}\left[1+C\left(\frac{S_i-E}{B}\right)^C\right]F(S_i)$$

$$Z(S_i) = \frac{1}{N_{mi}} - \Phi(S_i)$$

$$J_k = \sum_i [Z(S_i)]^2$$

3. The δA, δB, δC, δE values are calculated by the following matrix equation:

$$\begin{bmatrix} \delta A \\ \delta B \\ \delta C \\ \delta E \end{bmatrix} = \begin{bmatrix} \Phi'_A(S_1)\Phi'_B(S_1)\Phi'_C(S_1)\Phi'_E(S_1) \\ \Phi'_A(S_2)\Phi'_B(S_2)\Phi'_C(S_2)\Phi'_E(S_2) \\ \Phi'_A(S_3)\Phi'_B(S_3)\Phi'_C(S_3)\Phi'_E(S_3) \\ \Phi'_A(S_4)\Phi'_B(S_4)\Phi'_C(S_4)\Phi'_E(S_4) \end{bmatrix}^{-1} \begin{bmatrix} Z(S_1) \\ Z(S_2) \\ Z(S_3) \\ Z(S_4) \end{bmatrix}$$

4. Calculate the newest set of values

$$A_k = A_{k-1} + \delta A \quad B_k = B_{k-1} + \delta B \quad C_k = C_{k-1} + \delta C \quad E_k = E_{k-1} + \delta E$$

The best fitting curve is obtained when: $\frac{|J_k - J_{k-1}|}{J_1} < 0.001$ obtained.
If the best fitting curve is not obtained, restart in 2.

3. **Using software R**

The software R is a free software which can be downloaded from the web.

3.1. **Download and Install R**

Precompiled binary distributions of the base system and contributed packages can be downloaded from the following sites:

- Download R for Windows
- Download R for Linux
- Download R for (Mac) OS X.

  Note If you use Linux: R is part of many Linux distributions, so check with your Linux package management system.

  For Windows you will access to the following list of subdirectories:

Subdirectory	Comments
base	Binaries for base distribution (managed by Duncan Murdoch) What is needed to install R for the first time
contrib	Binaries of contributed CRAN packages for R >= 2.11.x (managed by Uwe Ligges) There is also information on third party software available for CRAN Windows services and corresponding environment and make variables

(continued)

(continued)

Subdirectory	Comments
old contrib	Binaries of contributed CRAN packages for outdated versions of R < 2.11.x (managed by Uwe Ligges)
Rtools	Tools to build R and R packages (managed by Duncan Murdoch) What is needed to build your own packages on Windows, or to build R itself

Select "base", then:

DownloadR 3.3.2 for Windows (62 megabytes, 32/64 bit)

and follow the instructions.

3.2. **Calculation tools**

Mean S–N Curve

The values of *A*, *B*, *C*, *E* are obtained by solving the equation:

$$N_i = \frac{A}{\Phi(N_i) - E} \exp\left(-\left(\frac{\Phi(N_i) - E}{B}\right)^C\right)$$

The program uses the "R" function "unirout" which calculates the root of a nonlinear equation.

Then to obtain the Bastenaire curve the program uses the "R" function "nls" which determines the nonlinear (weighted) least squares estimates of the parameters of a nonlinear model.

A "R" program (script) is written: Bastenaire_iiw-XIII-wg1.R (see example).

It takes the data from the file: Bastenaire_iiw-XIII-wg1.dat and provides the results in the file: Bastenaire_iiw-XIII-wg1-results.txt.

General Procedure

After downloading R software and installation the procedure is the following:

1. Generation of the data file Bastenaire_iiw-wg1.dat

   The data file can be generated with software such as Word pad with the following format. The structure is to be as follows:
   # Calculation title
   # Date
   # Stress range Nb of cycles Failure (F = failed; NF = non-failed)
   ΔS_1 N_1 F
   ΔS_1 N_2 F
   ....
   ΔS_j N_i NF
   ΔS_j N_{i+1} F
   ......

The data are given in blocks of ΔS in decreasing or increasing order. No order is required for N.
Then the file is stored in the working directory with the indicated name.

2. Running "R" software

Open "R" software.
Select the working directories where the data are and where the results will be given (Files/Change current directory).
Start the running (Files/Source R code).

3. Reading of the results

In the working directory open the file Bastenaire_iiw-XIII-wg1-results.txt.

3.3 **Specific functions and data in "Bastenaire_iiw-wg1.R"**

Two specific functions have been written corresponding to the Bastenaire formula.

Estimated Cycles Number at Failure Medians

For each stress level *i* there are *j* tests performed.

When more than half of the number of samples *j* at a given stress range level *i* are failed, the median is obtained as follows:

- For each ΔS_i the data N_{ji} are ordered following the increasing number of cycles at failure (N_{1i}, …, N_{ki}, …, N_{ni}).
- The median is so that:
  if n is odd $N_{mi} = N_{ki}$ $\quad k = \frac{n+1}{2}$
  if n is even $N_{mi} = \frac{N_{ki}+N_{k(i+1)}}{2}$ $\quad k = \frac{n}{2}$
  Note If the number of failed specimens at a given stress range level is less than half of the tested specimens the median cannot be correctly estimated. The corresponding stress range level is not taken into account for the best fitting curve determination excepted for the parameter *E*.

Initial Values of Parameters and Constraints

To allow the curve best fit procedure the parameters *A*, *B*, *C* and *E* are to be given an initial value. The following values can be generally chosen:

$A_0 = 1.0E7 \quad B_0 = 100 \quad C_0 = 1.0$

$E_0 = \Delta S_{\min} - 1$ (MPa) $\Delta S_{\min}$ is the lower level with at least one failed specimen

Steps of Calculation

1. For the step k (first step $k = 0$), for each N_i calculate $\Phi_k(N_i)$ from the equation ("R" function "unirout"):

$$N_i = \frac{A_k}{\Phi_k(N_i) - E_k} \exp\left(-\left(\frac{\Phi_k(N_i) - E_k}{B_k}\right)^{C_k}\right)$$

Determine the $A_{k+1}, B_{k+1}, C_{k+1}, E_{k+1}$ with the "R" function "nls" which determines the nonlinear (weighted) least squares estimates of the parameters of a nonlinear model.

2. Calculate the distance J_k (step k) between the ΔS_i and $\Phi_k(N_i)$

$$J_k = \sum_i [\Delta S_i - \Phi_k(N_i)]^2$$

3. Verify if J has converged to a minimum

If $\frac{|J_k - J_{k-1}|}{J_1} < 0.001$ the best fitting curve is obtained
If not follow

4. Reset the parameters A, B, C, E so that:

$$A = A_{k+1} B = B_{k+1} C = C_{k+1}, E = E_{k+1}$$

and apply the constraints:

$$A > 0 B > 0 C > 0 E < E_k \text{ and } > 0$$

5. Restart in 1.

4. **Comments**

The number of stress range levels with more than half of the number of samples failed must be equal or greater to 4.

The obtained Bastenaire curve is valid in the region of the test data points. The curve outside this region, extrapolation for lower N ($N <$ minimum N of the data sets), may be not correct.

Example

1. **Data**

i	1	2	3	4
ΔS (MPa)	1 010	596	394	215
j	N	N	N	N
1	1580	2900	14,000	885,000
2	585	6700	9000	180,000
3	450	4500	25,000	560,000
4	1100		40,000	1,100,000
5	640			2,000,000 (nf)
6				2,000,000 (nf)

2. Use of Excel—Case 1

Mean N Values

The mean value of N_{mi} for ΔS_i is:
$\log(N_{mi}) = \frac{1}{n_i} \sum_{j=1}^{n} \log(N_{ji})$ where n_i is the number of data at level ΔS_i.

I	1		2		3	
J	N	$\log(N_{j1})$	N	$\log(N_{j2})$	N	$\log(N_{j3})$
1	450	2.6532	2900	3.4624	9000	3.9542
2	585	2.7672	4500	3.6532	14,000	4.1461
3	640	2.8062	6700	3.8261	25,000	4.3979
4	1100	3.0414			40,000	4.6021
5	1580	3.1987				
$\log(N_{mi})$		2.8933		3.6472		4.2751
N_{mi}		782		4438		18,841

For ΔS_4 the level presents non-crack samples. The mean is determined with the log-log probability graph:

The probabilities p_{ji} for each N_{ji} is given by:

$$p_{ji} = \frac{j - 03}{n_i + 0.4}$$

j	N	$\log(N_{j4})$	p_{ij}	$\Phi^{-1}(p_{ij})$
1	180,000	5.2553	0.109375	– 1.2299
2	560,000	5.7482	0.265625	– 0.6261
3	885,000	5.9469	0.421875	– 0.1971
4	1,100,000	6.0414	0.578125	0.1971
5	2,000,000	6.3010	0.734375	0.6261
6	2,000,000	6.3010	0.890625	1.2299
$\log(N_{mi})$		5.9325		
N_{mi}		856,001		

$$\log(N_{mi}) = 12.616/2.1266$$

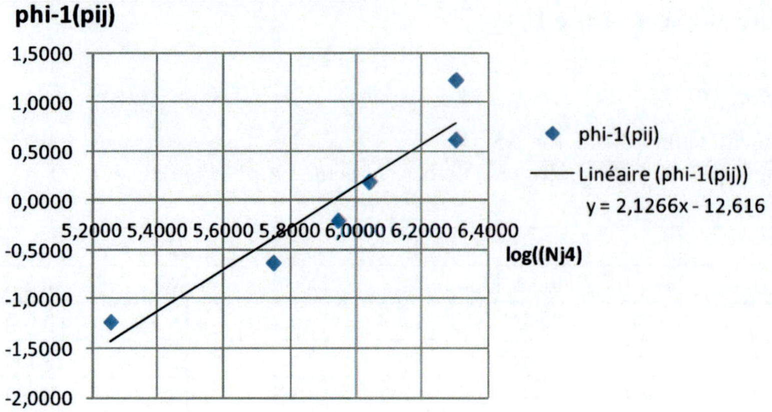

Mean S–N Curve

The estimation of the mean curve is performed by iteration considering a modification of the Bastenaire formula as follows:

$$\frac{1}{N} = \frac{S-E}{A} \exp\left(\left(\frac{S-E}{B}\right)^{C}\right) = \Phi(S)$$

The following values are considered:

$$A_0 = 1.0E06 \quad B_0 = 500 \quad C_0 = 1.0 \quad E_0 = 214$$

The corresponding Bastenaire curve is the following:

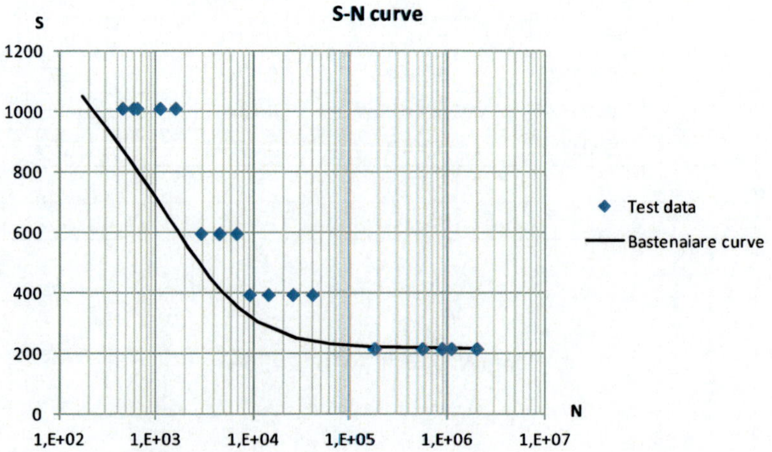

For each S_i level the following equation is solved (see procedure):

$$\begin{bmatrix} \delta A \\ \delta B \\ \delta C \\ \delta E \end{bmatrix} = \begin{bmatrix} \Phi'_A(S_1)\Phi'_B(S_1)\Phi'_C(S_1)\Phi'_E(S_1) \\ \Phi'_A(S_2)\Phi'_B(S_2)\Phi'_C(S_2)\Phi'_E(S_2) \\ \Phi'_A(S_3)\Phi'_B(S_3)\Phi'_C(S_3)\Phi'_E(S_3) \\ \Phi'_A(S_4)\Phi'_B(S_4)\Phi'_C(S_4)\Phi'_E(S_4) \end{bmatrix}^{-1} \begin{bmatrix} Z(S_1) \\ Z(S_2) \\ Z(S_3) \\ Z(S_4) \end{bmatrix}$$

and the following formula applied:

$$\Phi(S) = \Phi'_A\delta A + \Phi'_B\delta B + \Phi'_C\delta C + \Phi'_E\delta E$$

The following results are obtained

Step *k*	*A*	*B*	*C*	*E*	$\lvert J_k - J_{k-i}\rvert / J_1$
0	1.00E+06	500	1	214	
1	1.00E+06	5.00E+02	1.00E+00	2.14E+02	
2	1.9459E+06	379.6869	0.8461	212.8928	0.72242
3	3.7300E+06	234.3627	0.6902	210.8733	0.15444
4	6.9341E+06	124.4814	0.5761	207.6921	0.01407
5	1.1982E+07	75.3647	0.5107	204.0184	0.05033
6	1.8189E+07	55.5666	0.4757	201.071126	0.04568
7	2.3055E+07	48.5911	0.4605	199.5664	0.01216
8	**2.4737E+07**	**47.1207**	**0.4568**	**199.2038**	**0.00088**
9	2.4872E+07	47.0298	0.4566	199.1815	0.00001
10	2.4873E+07	47.0293	0.4566	199.1814	0.00000

Step *k*	Delta *A*	Delta *B*	delta *C*	Delta *E*
0	0	0	0	0
1	9.4588E+05	– 1.2031E+02	– 1.5392E–01	– 1.1072E+00
2	1.7841E+06	– 1.4532E+02	– 1.5593E–01	– 2.0196E+00
3	3.2041E+06	– 1.0988E+02	– 1.1405E–01	– 3.1812E+00
4	5.0482E+06	– 4.9117E+01	– 6.5421E–02	– 3.6737E+00
5	6.2066E+06	– 1.9798E+01	– 3.4947E–02	– 2.9473E+00
6	4.8657E+06	– 6.9755E+00	– 1.5249E–02	– 1.5048E+00
7	1.6819E+06	– 1.4704E+00	– 3.6772E–03	– 3.6262E–01
8	**1.3549E+05**	**– 9.0902E–02**	**– 2.4346E–04**	**– 2.2248E–02**
9	7.6824E+02	– 4.4776E–04	– 1.2376E–06	– 1.0269E–04

(continued)

(continued)

Step k	Delta A	Delta B	delta C	Delta E
10	2.4393E−02	− 1.3043E−08	− 3.6609E−11	− 2.7614E−09

The best fitting Bastenaire curve is given by:

A	B	C	E
2.474E+07	47.12	0.47	199.2

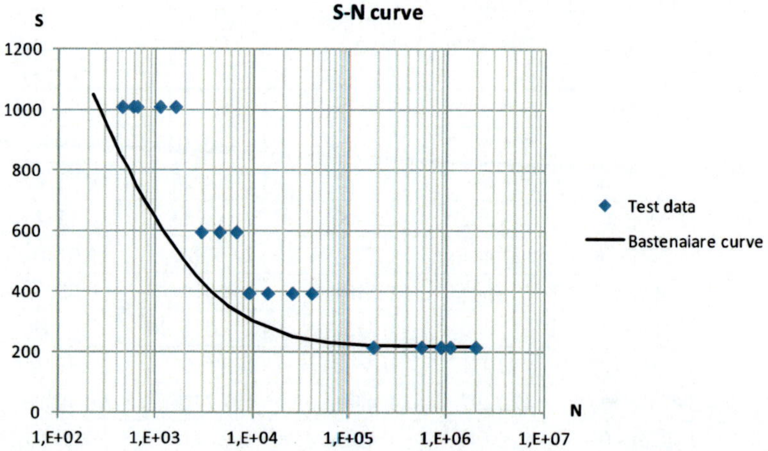

3. **Use of Excel—Case 2**

Another calculation has been performed with different initial values:

$$A_0 = 1.0\,E06 \quad B_0 = 600 \quad C_0 = 2.0 \quad E_0 = 214$$

The corresponding Bastenaire curve is the following:

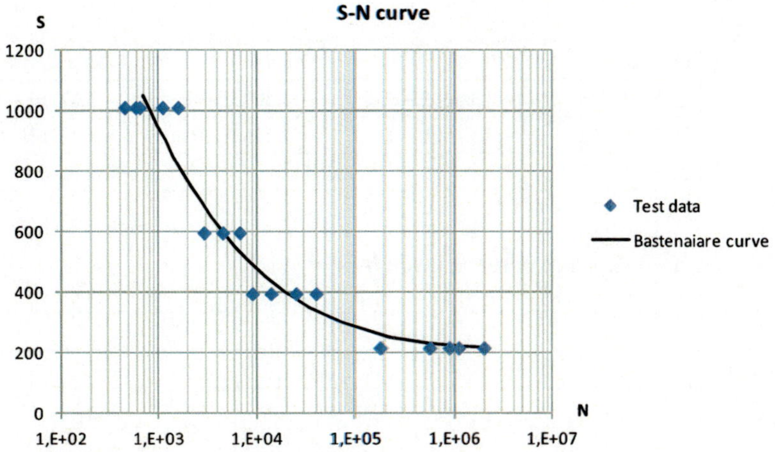

The iteration results are the following:

Step k	A	B	C	E	$\lvert J_k - J_{k-i}\rvert / J_1$
0	1.0000E+06	600	2	214	
1	1.0000E+06	600.000	2.000	214.000	
2	1.8300E+06	527.169	1.344	213.002	0.86807
3	3.3294E+06	368.688	0.819	211.231	0.12300
4	6.2352E+06	124.989	0.581	208.115	0.11511
5	1.0919E+07	81.523	0.520	204.635	0.07867
6	1.7050E+07	58.006	0.481	201.499	0.03309
7	2.2388E+07	49.3094	0.4622	199.7316	0.01120
8	2.4619E+07	47.2069	0.4570	199.2246	0.00107
9	**2.4870E+07**	**47.0309**	**0.4566**	**199.1818**	**0.00001**
10	2.4873E+07	47.0293	0.4566	199.1814	0.00000

Step k	Delta A	Delta B	Delta C	Delta E
0	0	0	0	0
1	8.3004E+05	− 7.2831E+01	− 6.5649E−01	−9.9824E−01
2	1.4994E+06	− 1.5848E+02	− 5.2475E−01	−1.7706E+00
3	2.9058E+06	− 2.4370E+02	− 2.3755E−01	−3.1158E+00
4	4.6843E+06	− 4.3467E+01	− 6.0871E−02	−3.4801E+00
5	6.1303E+06	− 2.3517E+01	− 3.9721E−02	−3.1361E+00
6	5.3377E+06	− 8.6967E+00	− 1.8415E−02	−1.7676E+00
7	2.2310E+06	− 2.1026E+00	− 5.1664E−03	−5.0693E−01
8	2.5160E+05	− 1.7596E−01	−4.6756E−04	−4.2873E−02

(continued)

(continued)

Step k	Delta A	Delta B	Delta C	Delta E
9	**2.6764E+03**	**− 1.6006E−03**	**− 4.4095E−06**	**−3.6760E−04**
10	2.9633E−01	− 1.6150E−07	− 4.5262E−10	−3.4276E−08
11	0	0	0	0

The best fitting Bastenaire curve is given by:

A	B	C	E
2.487E+07	47.03	0.46	199.2

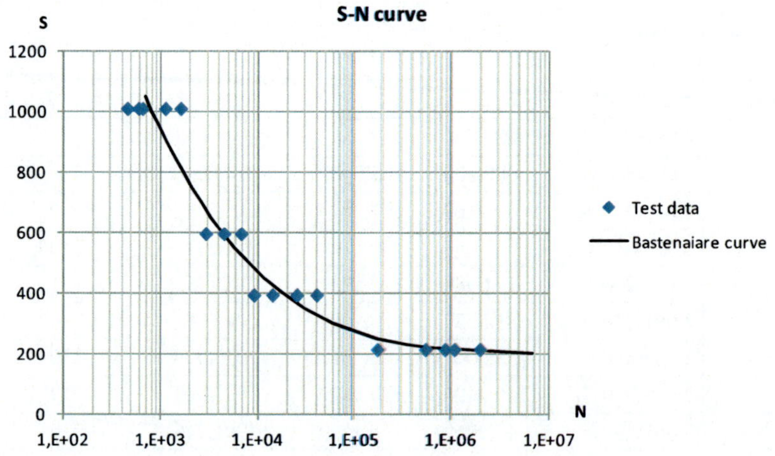

5. **R Software**

Estimated Medians

if n is odd $k = \frac{n+1}{2}$ $\quad N_{mi} = N_{ki}$

if n is even $k = \frac{n}{2}$ $\quad N_{mi} = \frac{N_{ki} + N_{(k+1)i}}{2}$

	N	N	N	N
j/i	1	2	3	4
2	450	2900	9000	180,000
3	585	4500	14,000	560,000
4	640	6700	25,000	885,000
5	1100		40,000	1,100,000
6	1580			2,000,000 (nf)

(continued)

(continued)

	N	N	N	N
7				2,000,000 (nf)
N_m	640	4500	19,500	992,500

Mean Curve Estimation

The following initial values and constraints of the Bastenaire curve coefficient are used:

$$A_0 = 1.0E + 07 \quad B_0 = 100.0 \quad C_0 = 1.0 \quad E_0 = \min(\Delta S) - 1$$
$$A > 1.0E + 06 \quad B > 10.0 \quad C > 0.1 \quad E > 20.0$$

The calculations are performed using the software R Bastenaire_iiw-XIII-wg1.R. The data are in the file Bastenaire_iiw-XIII-wg1.dat.

# Bastenaire curve—test			
# 22/02/2011			
#	Stress rang	Nb of cycles	
	1010.0	1580.0	F
	1010.0	585.0	F
	1010.0	450.0	F
	1010.0	1100.0	F
	1010.0	640.0	F
	596.0	2900.0	F
	596.0	6700.0	F
	596.0	4500.0	F
	394.0	14,000.0	F
	394.0	9000.0	F
	394.0	25,000.0	F
	394.0	40,000.0	F
	215.0	885,000.0	F
	215.0	180,000.0	F
	215.0	560,000.0	F
	215.0	1,100,000.0	F
	215.0	2,000,000.0	NF
	215.0	2,000,000.0	NF

Two specific functions have been written to solve the Bastenaire formula:

$$N_i = \frac{A}{\Phi(N_i) - E} \exp\left(-\left(\frac{\Phi(N_i) - E}{B}\right)^C\right)$$

using “R” function “unirout” which calculates the root of a nonlinear equation and “R” function “nls” to obtain the Bastenaire curve.

Function: f_Bastenaire for N_i
This function provides N_i for $x = \Phi(N_i)$

```
f_Bastenaire<- function(A B C E x)
{
f_Bastenaire<-(A*(1.0/(x-E))*exp(-((x-E)/B)^C))
return(f_Bastenaire)
}
```

Function: Phi_Bastenaire
This function provides x of the nonlinear Bastenaire equation for a given N ($N = N_i$)

```
Phi_Bastenaire<-function(A B C E N)
{
SB<-numeric(length(N))
for (i in 1:length(N))
{
res_x<- uniroot(function(x) f_Bastenaire(A B C E x)-N[i] interval=c(0.0
1.0E+50))
SB[i] <-res_x$root
}
Return(SB)
}
```

To estimate the values of A B C E so that $\sum(\Delta S - \Phi(N_m))^2$ is minimum “nls” is applied with the following parameters:

```
res3<-nls(S~Phi_Bastenaire(A B C E N)
    start=c(A=A0 B=B0 C=C0 E=E0)
    data=data S=data$S N=data$N
    algorithm="port"
    lower=c(A=1.0E06 B=10.0 C=0.1 E=20.0)
    control=list(maxiter=1000))
```

Results

The results are in the file Bastenaire_iiw-XIII-wg1-results.txt

From f_Bastenaire function the following values of $\Phi(N_m)$ are obtained from the initial values of A, B, C, E and compared with ΔS

i	N_m	$\Phi(N_m)$	ΔS
1	992,500	223.19	215.0
2	19,800	348.12	384.0
3	4500	441.79	596.0
4	640	587.40	1010.0

As $\Phi(N_m)$ is different of ΔS the set of parameter (A B C E) need to be improved by the best fitting procedure.

After some iteration the obtained results are the following:

$$A = 10{,}925{,}513 \quad B = 167.45 \quad C = 0.71 \quad E = 205.35$$

The final values $\Phi(N_m)$ are the following:

i	N_m	$\Phi(N_m)$	ΔS
1	992,500	215.0	215.0
2	19,800	384.0	384.0
3	4500	596.0	596.0
4	640	1010.0	1010.0

Which shows that the best fitting curve is exactly on the test median points.

The data and formula graph is the following:

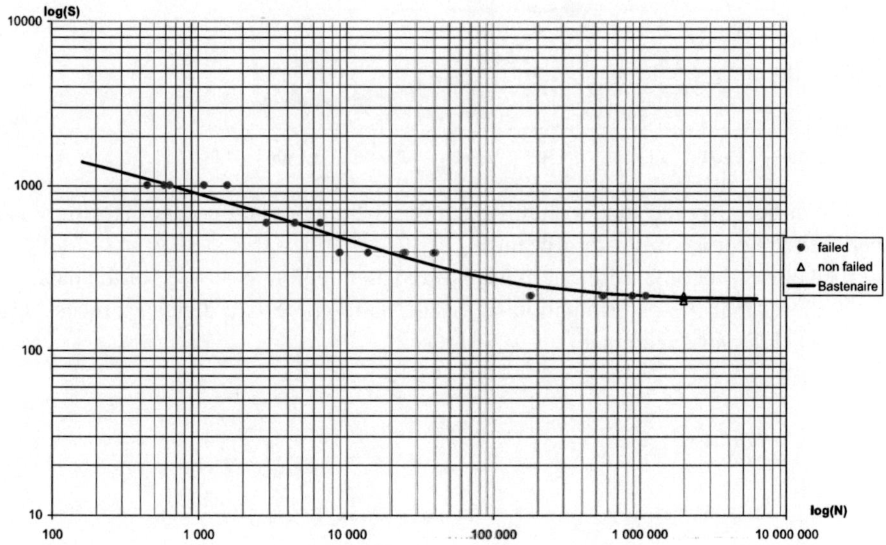

6. **Comments**

The two applications on the same test data provide the following Bastenaire parameters:

	Excel 1	Excel 2	R software
A	2.5776E+07	2.4873E+07	1.0925E+07
B	44.8511	47.0293	167.45
C	0.4533	0.4566	0.71
E	207.0787	199.1814	205.35

Excel 1 and 2 results are similar but R software looks different. The corresponding curves are the followings:

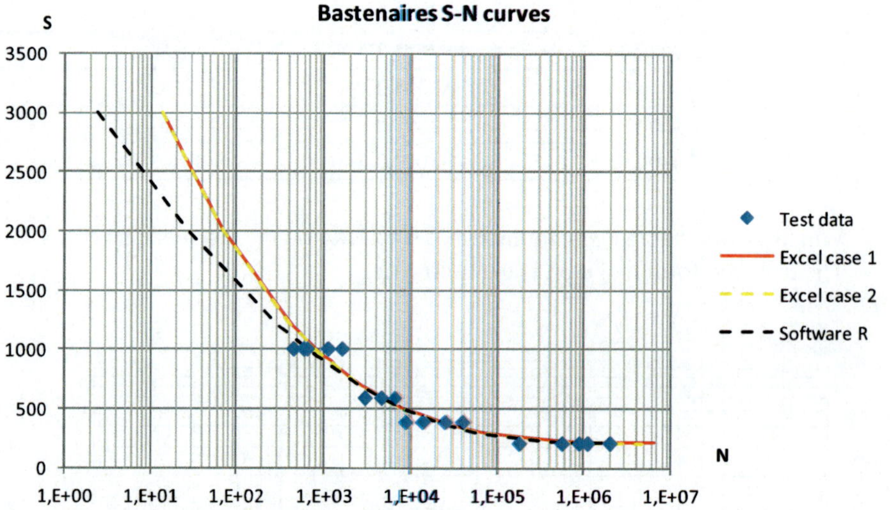

It can be observed that within the region of the test data points the difference between the curves is not significant.

The curves become different in the extrapolated region, $N < N_{\min}$ of the data.

It shows that the determination of a Bastenaire curve is a delicate process and requires skill and experience.

2.15 Sheet 14: Are Two Experimental Design S–N Curves Statistically Equivalent?

General

1. **Context**

The problem is raised when two data sets have been collected under different conditions but are expected to give comparable fatigue performance.

The question is to determine if there is statistically a significant difference between the two data sets.

2. **Principle**

Whether two experimental design S–N curves are or not statistically equivalent can be determined by using a hypothesis test.

As the design curve is established by adopting a certain number of standard deviations below the mean S–N curve, the question can be solved by determining the statistical equivalence of the mean S–N curve (constant and slope) and standard deviation.

A lower significance level is often used for each individual hypothesis test. For instance a significance level of 1.7% for each individual test would roughly correspond to a 5% significance level for the composite hypothesis.

3. **Condition of application**

The S–N curve equation is $m \log(S) + \log(N) = \log(K)$.

Design curves will only be applied to values of $\log(S)$ that are not far from the mean value.

The necessary data to perform the equivalence verification are the following:

- Two data sets (S_1, n_1) values and (S_2, n_2) values
- All samples have failed samples.
- The data sets follow a normal distribution, if not, the number of n_1 and n_2 shall be greater or equal to 30.

Procedure

1. **Criteria**

The applied tests are:

- two-sided Student's t distribution
- one-sided Snedecor F variable.

The selected level of risk in probability to reject a correct hypothesis is $\alpha\%$.

2. **Data and formula**

Determination of the Two Mean S–N Curves

The correct values of the constants K and m are obtained from the following two equations by using the maximum likelihood method:

	Sample $j = 1$	Sample $j = 2$
Sample size—n_i	…….	…….
Sum of N values—$\sum_j \log(N_{ij})$	…….	…….
Mean N values—$\overline{\log(N_i)} = \frac{\sum_j N_{ij}}{n_i}$	…….	…….
Sum of S values—$\sum_j \log(S_{ij})$	……	……
Mean S values—$\overline{\log(S_i)} = \frac{\sum_j S_{ij}}{n_i}$	……	……
Slopes m_i estimated or fixed from return experience: if estimated $m_i = -\frac{\sum_j (\log(S_{ij}) \log(N_{ij})) - \overline{\log(S_i) \log(N_i)}}{\sum_j (\log(S_{ij}))^2 - \overline{\log(S_i)}^2}$	……	……
Estimated values of $\log(K_i) = \overline{\log(N_i)} + m_i \overline{\log(S_i)}$	……	……

Note The calculations and graph drawing can be done with Excel.

Degree of freedom f_i if estimated: $f_i = n_i - 2$/if fixed $f_i = n_i - 1$	……	…….
Calculate the estimated values $\log\left(\tilde{N}_{ij}\right) = \log(K_i) - m_i \log(S_{ij})$		
Estimated variances $v_i = \frac{\sum_j \left(\log(N_{ij}) - \log\left(\tilde{N}_{ij}\right)\right)^2}{f_i}$	……	……
Common variance of the two samples $v_e = \frac{f_1 v_1 + f_2 v_2}{f_1 + f_2}$	……	
Degree of freedom of the two samples f_e if m_i are estimated $f_e = n_1 + n_2 - 4$/if m_i are fixed $f_e = n_1 + n_2 - 2$	……	
Two-sided Student's t distribution $t(1 - \alpha, f_e)$	……	

Note There is various definition of the function $t(\beta, \nu)$. The considered definition in the present sheet is so that: $t(0.95, 10) = 2.228$.

Testing That the Slopes of the Two S–N Curves Are Consistent

Difference of estimated slopes—$T = \|m_1 - m_2\|$	……
Criteria $\mathrm{Cr} = T - t(1 - \alpha, f_e) \sqrt{\left[\frac{1}{\sum_j (\log(S_{1j}) - \overline{\log(S_1)})^2} + \frac{1}{\sum_j (\log(S_{2j}) - \overline{\log(S_2)})^2}\right] v_e}$	……

Conclusion:

If $\mathrm{Cr} > 0$ the hypothesis of consistency has to be rejected.

Test That the Constants K of the Two S–N Are Consistent

Difference of estimated constant $X = \lvert\log(K_1) - \log(K_2)\rvert$	……
Criteria	……
$\mathrm{Cr} = X - t(1-\alpha, f_e)\sqrt{\left[\frac{1}{n_1} + \frac{1}{n_2} + \frac{\left(\overline{\log(S_1)}\right)^2}{\sum_j \left(\log(S_{1j}) - \overline{\log(S_1)}\right)^2} + \frac{\left(\overline{\log(S_2)}\right)^2}{\sum_j \left(\log(S_{2j}) - \overline{\log(S_2)}\right)^2}\right] v_e}$	

Conclusion:

If $\mathrm{Cr} > 0$ the hypothesis of consistency has to be rejected.

Verification That the Two Variances Are the Consistence

Variances ordered	if $v_1 > v_2$ $v_n = v_1$	if $v_1 < v_2$ $v_n = v_2$	$v_n =$ ……
	if $v_1 > v_2$ $v_d = v_2$	if $v_1 < v_2$ $v_d = v_1$	$v_d =$ ……
Degree of freedom	if $v_1 > v_2$ $f_n = f_1$	if $v_1 < v_2$ $f_n = f_2$	$f_n =$ ……
	if $v_1 > v_2$ $f_d = v_2$	if $v_1 < v_2$ $f_d = v_1$	$f_d =$ ……
Variance ratio	$R = \frac{v_n}{v_d}$		…….
One-sided Snedecor F-distribution $F(1-\alpha, f_n, f_d)$			……

Note The considered definition of the function $F(p, f_1, f_2)$ in the present sheet is so that $F(0.95, 10, 15) = 2.5437$.

Criteria $\mathrm{Cr} = R - F(1-\alpha, f_n, f_d)$	…….

Conclusion:

If $\mathrm{Cr} > 0$ the hypothesis of consistency has to be rejected.

3. **Conclusion**

If one of above hypothesis is rejected the composite hypothesis cannot be accepted.

If all of above hypothesis have no reason to be rejected, the composite hypothesis can be accepted.

Example

1. **Data and formula**

Samples data

No. 1, $n_1 = 14$		No. 2, $n_2 = 15$	
Stress range	Cycles to failure	Stress range	Cycles to failure
147	521,382	136	1,133,000
96	1,879,752	74	3,675,000
250	115,816	53	22,082,998

(continued)

(continued)

No. 1, $n_1 = 14$		No. 2, $n_2 = 15$	
Stress range	Cycles to failure	Stress range	Cycles to failure
61	10,204,041	53	7,154,785
57	14,910,395	136	664,000
57	10,646,018	75	3,816,199
96	1,475,769	136	787,894
250	112,910	136	689,973
74	6,766,000	147	958,318
74	6,195,271	53	13,475,000
74	4,310,000	53	20,511,538
136	1,015,824	265	95,982
136	1,580,669	54	19,048,838
53	11,541,520	74	9,232,000
		176	605,721

2. **Calculation**

The risk level is fixed to $\alpha = 1.7\%$ for each parameters, i.e. 5% for the composite hypothesis of consistency.

Determination of the two mean S–N curves

The correct values of the constants K and m are obtained from the following two equations by using the maximum likelihood method:

	Sample $j = 1$	Sample $j = 2$
Sample size—n_i	14	15
Sum of N values—$\sum_j \log(N_{ij})$	89.011	96.390
Mean N values—$\overline{\log(N_i)} = \frac{\sum_j N_{ij}}{n_i}$	6.358	6.426
Sum of S values$\sum_j \log(S_{ij})$	27.824	29.613
Mean S values—$\overline{\log(S_i)} = \frac{\sum_j S_{ij}}{n_i}$	1.987	1.974
Slopes m_i estimated:$-\frac{\sum_j(\log(S_{ij})\log(N_{ij}))-\overline{\log(S_i)}\overline{\log(N_i}}{\sum_j(\log(S_{ij}))^2-\overline{\log(S_i)}^2}$	3.06	3.01
Estimated values of $\log(K_i) = \overline{\log(N_i)} + m_i\overline{\log(S_i)}$	12.444	12.365

Note The calculations and graph drawing can be done with Excel

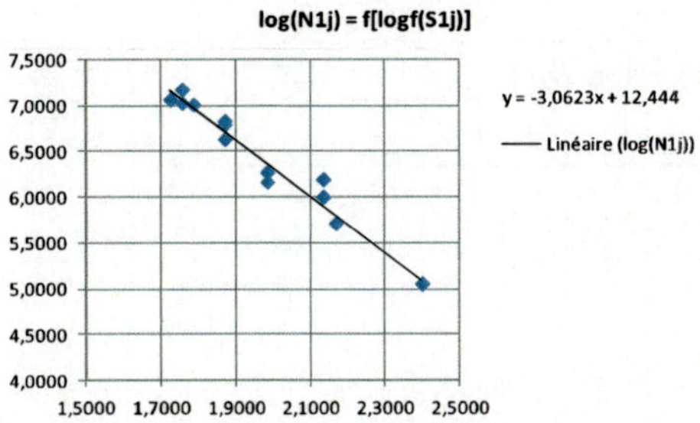

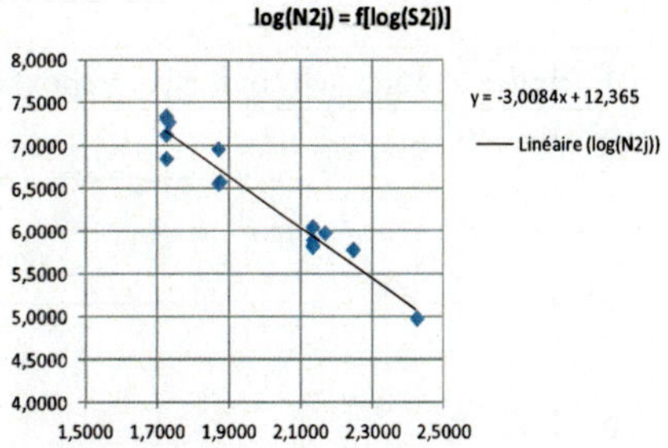

Degree of freedom, m_i estimated—$f_i = n_i - 2$	12	13
Calculate the estimated values $\log\left(\tilde{N}_{ij}\right) = \log(K_i) - m_i \log\left(S_{ij}\right)$		
Estimated variances $v_i = \frac{\sum_j \left(\log(N_{ij}) - \log\left(\tilde{N}_{ij}\right)\right)^2}{f_i}$	0.0171	0.0283
Common variance of the two samples $v_e = \frac{f_1 v_1 + f_2 v_2}{f_1 + f_2}$	0.0198	
Degree of freedom of the two samples, mi estimated, $f_e = n_1 + n_2 - 4$	25	
Risk level is fixed to α	0.017	
Two-sided Student's t distribution $t(1 - \alpha, f_e)$	2.557	

Note There is various definition of the function $t(\beta, \nu)$. The considered definition in the present sheet is so that: $t(0.95, 10) = 2.228$

Testing That the Slopes of the Two S–N Curves Are Consistent

Difference of estimated slopes—$T = \lvert m_1 - m_2 \rvert$	0.054
Criteria $\mathrm{Cr} = T - t(1 - \alpha, f_e)\sqrt{\left[\frac{1}{\sum_j \left(\log(S_{1j}) - \overline{\log(S_1)}\right)^2} + \frac{1}{\sum_j \left(\log(S_{2j}) - \overline{\log(S_2)}\right)^2}\right] v_e}$	− 0.596

Conclusion:

Cr < 0 the hypothesis of consistency has no reason to be rejected.

Test That the Constants K of the Two S–N Are Consistent

Difference of estimated constant $X = \lvert\log(K_1) - \log(K_2)\rvert$	0.0789
Criteria	– 1.216
$Cr = X - t(1-\alpha, f_e)\sqrt{\left[\frac{1}{n_1} + \frac{1}{n_2} + \frac{\left(\overline{\log(S_1)}\right)^2}{\sum_j \left(\log(S_{1j}) - \overline{\log(S_1)}\right)^2} + \frac{\left(\overline{\log(S_2)}\right)^2}{\sum_j \left(\log(S_{2j}) - \overline{\log(S_2)}\right)^2}\right] v_e}$	

Conclusion:

Cr < 0 the hypothesis of consistency has no reason to be rejected.

Verification That the Two Variances Are the Consistence

Variances ordered	if $v_1 > v_2$ $v_n = v_1$	if $v_1 < v_2$ $v_n = v_2$	$v_n = 0.0283$
	if $v_1 > v_2$ $v_d = v_2$	if $v_1 < v_2$ $v_d = v_1$	$v_d = 0.0171$
Degree of freedom	if $v_1 > v_2$ $f_n = f_1$	if $v_1 < v_2$ $f_n = f_2$	$f_n = 13$
	if $v_1 > v_2$ $f_d = v_2$	if $v_1 < v_2$ $f_d = v_1$	$f_d = 12$
Variance ratio	$R = \frac{v_n}{v_d}$		1.658
Risk level is fixed to α			0.017
One-sided Snedecor F-distribution $F(1-\alpha, f_n, f_d)$			3.587

Note The considered definition of the function $F(p, f_1, f_2)$ in the present sheet is so that: $F(0.95, 10, 15) = 2.5437$

Criteria $Cr = R - F(1-\alpha, f_n, f_d)$	– 1.929

Conclusion:

Cr < 0 the hypothesis of consistency has no reason to be rejected.

3. **Conclusion**

The above three individual hypothesises with a significance level 1.7% are accepted.

We can draw the conclusion that the composite hypothesis is acceptable with a significance level 5%.

2.16 Sheet 15: How to Determine the Degree of Improvement Produced by a Post-weld Treatment Process?

General and Procedure

1. **Context**

A post-weld treatment is implemented to improve the weld toe profile and remove welding residual stresses in order to improve the fatigue behaviour.

Commonly used methods are:

- improving weld toe profile, grinding
- improving both weld toe profile and welding residual stresses, tig or plasma refusion, wire bundle, hammer peening, shot peening and ultrasonic peening.

The magnitude of improvement depends primarily on the joint severity and base material.

2. **Principle**

The justification of an improvement with respect to as-welded joints requires fatigue tests of representative samples of treated joints and either a standard *S–N* curve or fatigue tests of as-welded samples.

The improvement requires determining the slope *S–N* curve parameters m and constant or fatigue class FAT (lifetime at 2 millions cycles).

The fatigue class FAT corresponds to a fatigue life with 95% survival probability.

3. **Method**

The two following cases are considered:

- as-welded *S–N* curve defined by a standard *S–N* curve
- as-welded fatigue life defined by sample fatigue test results.

As-Welded S–N Curve Defined by a Standard S–N Curve

The standard *S–N* curve is defined by its slope parameter m_{st} and K_{st} or FAT_{st}.

The post-welded fatigue test data (S_i, N_i) allow to determine a design *S–N* curve and the corresponding parameters m_{imp} and K_{imp} or FAT_{imp}.

As the improvement may have an influence on the slope parameter m, the data set (S_i, N_i) must be analysed with the sheets 10 or 12.

When m_{st} and m_{imp} are not significantly different, the data set (S_i, N_i) must be newly analysed with the sheets 9 or 11, m fixed equal to m_{st}.

When m_{st} and m_{imp} are significantly different the measurement of the improvement will require a miner sum calculation with a representative stress range distribution.

As-Welded Fatigue Life Defined by Sample Fatigue Test Results

Two data sets are available:

- as-welded fatigue test data (S_{1i}, N_{1i})
- post-welded fatigue test data (S_{2i}, N_{2i}).

To determine if there is an improvement or not it must be applied the sheet 14 which also provides the mean *S–N* curve parameters.

If the improvement can be considered significant, the post-welded design *S–N* curve can be determined with sheets 10 or 12.

Appendix 1: One-Sided Tolerance Limit Factors *k* Slope *m* Fixed

Degree of freedom, $f = n - 1$.

Proportion of non-failure $p = 0.95$ for one-sided or $p = 0.90$ for two-sided interval, $\Phi^{-1}(p) = 1.6649$.

Confidence $\alpha = 0.75$ for bilateral interval or $\alpha = 0.875$ for unilateral interval.

Sample size, *n*	Value of *k*
3	4.704088
4	3.620119
5	3.161453
6	2.903646
7	2.73611
8	2.617255
9	2.527849
10	2.457725
11	2.400978
12	2.35393
13	2.31416
14	2.280008
15	2.250294
16	2.224153
17	2.200938
18	2.180151
19	2.161406
20	2.144396
21	2.128874
22	2.114639
23	2.101525
24	2.089397
25	2.078138

(continued)

(continued)

Sample size, n	Value of k
26	2.067652
27	2.057855
28	2.048677
29	2.040056
30	2.031939
31	2.02428
32	2.017037
33	2.010176
34	2.003664
35	1.997473
36	1.991578
37	1.985957
38	1.980589
39	1.975456
40	1.970542
41	1.965832
42	1.961313
43	1.956971
44	1.952796
45	1.948778
46	1.944908
47	1.941175
48	1.937574
49	1.934096
50	1.930734
60	1.902342
70	1.880824
80	1.86381
90	1.849934
100	1.838343
110	1.828477
120	1.81995
130	1.812487
140	1.805884
150	1.79999
160	1.794687
170	1.789882

(continued)

(continued)

Sample size, n	Value of k
180	1.785504
190	1.781493
200	1.777801
210	1.774388
220	1.77122
230	1.76827
240	1.765514
250	1.762932

Appendix 2: One-Sided Tolerance Limit Factors *k* Slope *m* Estimated

Degree of freedom, $f = n - 2$.

Proportion of non-failure $p = 0.95$ for one-sided or $p = 0.90$ for two-sided interval, $\Phi^{-1}(p) = 1.6649$.

Confidence $\alpha = 0.75$ for bilateral interval or $\alpha = 0.875$ for unilateral interval.

n	$\frac{n^*}{n}$									
	0.01	0.02	0.03	0.04	0.05	0.06	0.07	0.08	0.09	0.10
3	19.89138098	16.03966818	14.4050842	13.47102153	12.85988632	12.42737127	12.10509526	11.85605136	11.65830186	11.49794609
4	10.98002589	8.694592369	7.701503509	7.12038487	6.730906871	6.448437287	6.232665972	6.061662306	5.922354895	5.806414435
5	8.764014751	6.937510027	6.139768729	5.670656979	5.354734319	5.124529775	4.947868181	4.807223269	4.692135756	4.595932347
6	7.663736962	6.08320572	5.391394314	4.983731152	4.708639721	4.507796235	4.353373094	4.230204235	4.129234063	4.044681746
7	6.970848479	5.552444762	4.930856159	4.564157168	4.316436988	4.135384644	3.996034213	3.884775309	3.793478845	3.716953899
8	6.480082701	5.180040237	4.609891777	4.273298621	4.045759206	3.879345163	3.751178065	3.648783264	3.564708642	3.494194675
9	6.107494142	4.899266256	4.369109431	4.05597183	3.844188166	3.689226132	3.569825748	3.47439339	3.396002362	3.330228154
10	5.811401473	4.677321734	4.179512086	3.885374915	3.686372999	3.540714599	3.428446288	3.338685941	3.264931358	3.203028788
11	5.568360277	4.495909395	4.02501645	3.746707379	3.558364601	3.420472963	3.314164688	3.229149025	3.159276695	3.100619122
12	5.36398873	4.343882569	3.895871719	3.631029644	3.451763422	3.320490706	3.219265624	3.138299713	3.071743513	3.015859785
13	5.188883583	4.213995981	3.785765949	3.532573442	3.361163953	3.235624247	3.138804718	3.061350806	2.997672099	2.944196703
14	5.036589246	4.101300515	3.690402816	3.447423412	3.282905487	3.162397083	3.069445803	2.995077068	2.93392735	2.882569593
15	4.902499603	4.002280247	3.606739528	3.372813294	3.214406879	3.098362165	3.008844144	2.937214751	2.878311415	2.828835379

n	$\frac{n^*}{n}$									
	0.01	0.02	0.03	0.04	0.05	0.06	0.07	0.08	0.09	0.10
16	4.783223916	3.914356746	3.532550645	3.306724207	3.153787694	3.041739466	2.955296408	2.886121305	2.829231256	2.781442226
17	4.676200865	3.835588766	3.466164656	3.24764265	3.09964044	2.991198451	2.907530908	2.840571718	2.78549992	2.739234884
18	4.579453201	3.764482134	3.406297723	3.194408103	3.050887402	2.945721573	2.864576133	2.799631033	2.746212277	2.701333086
19	4.491426043	3.699864925	3.351944796	3.146113425	3.006687232	2.90451532	2.825675206	2.762571476	2.710664316	2.667052647
20	4.410876942	3.6408029	3.302306023	3.102037667	2.966372035	2.866950398	2.790228409	2.728816838	2.67829904	2.635852601
21	4.33679917	3.586540664	3.256735668	3.06159951	2.92940383	2.832520371	2.757753493	2.697904106	2.648669159	2.607298792
22	4.268366916	3.536459762	3.214705791	3.024324216	2.895343687	2.80081239	2.727857609	2.669456347	2.621410758	2.581038179
23	4.204895355	3.490048221	3.175779908	2.989819686	2.863829398	2.771486024	2.700217039	2.643163147	2.596224305	2.556780325
24	4.145811021	3.446878004	3.139593523	2.957758802	2.834559022	2.744257684	2.674562278	2.618766216	2.572860691	2.534283798
25	4.090629477	3.406588059	3.105839509	2.927866168	2.807278548	2.71888895	2.650666881	2.596048645	2.551110816	2.513346027
26	4.038938242	3.368871377	3.074256972	2.899908036	2.781772525	2.695177715	2.628338989	2.574826744	2.530797677	2.493795605
27	3.990383564	3.333464991	3.04462265	2.873684522	2.757856833	2.67295136	2.607414818	2.55494377	2.5117703	2.475486384
29	3.901502518	3.268706158	2.990454883	2.825776021	2.71418402	2.632379479	2.569233503	2.518674122	2.477071732	2.442106671
30	3.860679314	3.238985385	2.965609351	2.80381208	2.694170299	2.61379367	2.55174862	2.502069712	2.461191004	2.426833507
31	3.82198714	3.210829336	2.94208022	2.783017963	2.675227331	2.596206166	2.535206253	2.486363255	2.446171644	2.412391029
32	3.785246686	3.184105396	2.919755321	2.763293553	2.657263132	2.579530925	2.51952493	2.471476923	2.431938813	2.398706905
33	3.75029911	3.158696184	2.898535434	2.744550335	2.640196414	2.563691939	2.504632702	2.457342008	2.41842646	2.38571731
34	3.717003125	3.134497376	2.878332423	2.726709719	2.623955035	2.548621773	2.49046576	2.443897592	2.40557604	2.373365683
35	3.685232575	3.111415889	2.859067694	2.709701657	2.608474717	2.534260367	2.476967286	2.431089447	2.393335446	2.3616017

n	$\frac{n^*}{n}$									
	0.01	0.02	0.03	0.04	0.05	0.06	0.07	0.08	0.09	0.10
36	3.654874412	3.089368371	2.840670905	2.693463481	2.593697975	2.520554023	2.464086498	2.418869117	2.381658133	2.350380415
37	3.625826993	3.068279927	2.823078885	2.677938934	2.579573224	2.507454569	2.45177785	2.407193154	2.370502375	2.339661543
38	3.597998637	3.048083047	2.806234716	2.663077348	2.566054017	2.494918646	2.440000359	2.396022471	2.35983064	2.329408859
39	3.571306405	3.028716694	2.790086961	2.648832953	2.553098411	2.482907108	2.428717033	2.385321797	2.349609068	2.319589684
40	3.545675056	3.010125528	2.774589005	2.635164279	2.540668417	2.471384509	2.417894384	2.375059206	2.339807018	2.310174455
41	3.521036149	2.992259238	2.759698485	2.622033656	2.528729534	2.460318666	2.407502013	2.365205726	2.330396687	2.301136348
42	3.497327273	2.97507197	2.745376806	2.609406774	2.517250347	2.449680282	2.397512254	2.355734993	2.32135278	2.292450966
43	3.474491379	2.958521829	2.731588721	2.597252307	2.506202183	2.439442623	2.387899866	2.346622958	2.312652228	2.284096061
44	3.452476205	2.942570456	2.718301964	2.585541589	2.495558808	2.429581234	2.378641766	2.337847632	2.304273941	2.276051297
45	3.431233765	2.927182644	2.705486935	2.574248331	2.485296171	2.4200737	2.369716796	2.329388867	2.296198597	2.268298044
46	3.410719916	2.912326021	2.693116425	2.56334837	2.47539217	2.410899426	2.361105525	2.321228158	2.288408452	2.260819199
47	3.390893965	2.897970758	2.681165371	2.552819457	2.465826458	2.402039458	2.35279007	2.313348478	2.280887184	2.253599029
48	3.371718333	2.88408932	2.669610642	2.542641064	2.456580267	2.393476314	2.34475394	2.305734128	2.273619745	2.246623031
49	3.353158256	2.870656242	2.658430855	2.532794218	2.447636252	2.385193842	2.336981902	2.298370611	2.266592239	2.239877817
50	3.335181515	2.857647934	2.647606206	2.52326135	2.438978357	2.377177093	2.329459859	2.29124451	2.259791809	2.233350998
60	3.181828458	2.746780612	2.555413798	2.44211762	2.36531912	2.309002751	2.265518071	2.230690604	2.202024784	2.177925564
70	3.063718004	2.661514279	2.484586811	2.379834799	2.308825112	2.256751629	2.216541557	2.184335575	2.157826495	2.135539656
80	2.969145667	2.593315665	2.427984362	2.330094922	2.263735261	2.215070362	2.17749127	2.147391732	2.122615831	2.101785583
90	2.891212502	2.537165383	2.381412597	2.289192183	2.226674133	2.180825458	2.145420293	2.117061401	2.093717808	2.07409138
100	2.82555375	2.489892929	2.342225666	2.254790979	2.195516181	2.152045234	2.11847562	2.091586527	2.069452442	2.050842633

n	$\frac{n^*}{n}$									
	0.01	0.02	0.03	0.04	0.05	0.06	0.07	0.08	0.09	0.10
110	2.76925538	2.449384339	2.308661157	2.225336812	2.168847711	2.127419144	2.095426279	2.06979981	2.048704795	2.030968389
120	2.720287813	2.414168862	2.279493904	2.199749783	2.145687203	2.106037757	2.075418453	2.050891944	2.030702171	2.013726702
130	2.677188297	2.383187475	2.253842325	2.17725328	2.125329153	2.087247668	2.057838941	2.034281904	2.014890008	1.99858525
140	2.638872408	2.355655681	2.231053758	2.157272641	2.107251731	2.070565769	2.042234514	2.019540378	2.00085867	1.985150923
150	2.604516905	2.330978337	2.210633332	2.139372322	2.091059588	2.055626165	2.028262011	2.006342401	1.988298162	1.973126307
160	2.573484059	2.308694695	2.192198105	2.123215417	2.076447008	2.042146004	2.015656222	1.994436888	1.976969017	1.9622817
170	2.545271167	2.288441758	2.175446498	2.108536697	2.063173387	2.029902731	2.004208522	1.983626367	1.966682941	1.952436518
180	2.519475893	2.269929143	2.160137353	2.09512413	2.051046432	2.01871852	1.993752254	1.973753121	1.957289571	1.943446581
190	2.495771898	2.252921423	2.146075216	2.082805918	2.039910374	2.008449352	1.984152468	1.964689462	1.948667196	1.935195198
200	2.473891333	2.237225442	2.133099775	2.07144119	2.029637513	1.998977183	1.975298572	1.956330765	1.940716101	1.927586782
210	2.453612018	2.22268104	2.121078139	2.060913194	2.020122035	1.990204217	1.967098962	1.948590382	1.933353714	1.920542188
220	2.434747897	2.20915414	2.109899092	2.051124242	2.011275419	1.982048655	1.959477025	1.941395855	1.926510998	1.913995252
230	2.417141826	2.196531537	2.099468749	2.041991886	2.003022969	1.974441481	1.952368132	1.934686067	1.920129725	1.907890176
240	2.40066004	2.184716895	2.089707235	2.033445992	1.995301153	1.967324006	1.945717332	1.928409059	1.914160398	1.902179533
250	2.385187867	2.173627651	2.080546107	2.025426478	1.988055553	1.960645963	1.939477563	1.922520337	1.908560637	1.896822718

n	$\frac{n^*}{n}$									
	0.1	0.2	0.3	0.4	0.5	0.6	0.7	0.8	0.9	1
3	11.49794609	10.78069861	10.57754202	10.50047661	10.46758731	10.45287358	10.44632275	10.44362243	10.44277947	10.44282967
4	5.806414435	5.221303607	4.996456732	4.877380646	4.803923551	4.754233906	4.71845433	4.691494637	4.670467401	4.653616277
5	4.595932347	4.101269132	3.903855254	3.796070119	3.727909006	3.680861622	3.646421936	3.620118985	3.599374069	3.582594411
6	4.044681746	3.60665821	3.429245336	3.331238	3.268667267	3.225139359	3.193068435	3.168440976	3.148927928	3.133082576
7	3.716953899	3.318910582	3.156408693	3.066073369	3.008104608	2.967606639	2.937661769	2.914597447	2.896275669	2.881364463
8	3.494194675	3.126487464	2.975625409	2.891430759	2.837228445	2.799260134	2.771122122	2.749407578	2.732129286	2.718046948
9	3.330228154	2.986641276	2.845196557	2.766044749	2.71497599	2.679136403	2.652534226	2.631977244	2.61560094	2.602240152
10	3.203028788	2.879257135	2.745640381	2.670722113	2.622306595	2.588282881	2.56299937	2.543441975	2.527848523	2.515116812
11	3.100619122	2.793524568	2.66655208	2.595252687	2.549118957	2.516665091	2.492526802	2.473841009	2.458932596	2.446753101
12	3.015859785	2.723065894	2.601827369	2.533667632	2.489522483	2.458442078	2.435309146	2.41739078	2.403087115	2.39139625
13	2.944196703	2.663849777	2.547626881	2.482224798	2.439832395	2.409966211	2.387724401	2.370487823	2.356722506	2.345467402
14	2.882569593	2.613189025	2.501402817	2.438448112	2.397615584	2.3688326	2.347387407	2.330761425	2.317479018	2.306615358
15	2.828835379	2.569215383	2.461390975	2.400627568	2.361194874	2.333385854	2.312658158	2.296582942	2.283736695	2.273226996
16	2.781442226	2.530584766	2.426326776	2.367540391	2.329373037	2.30244583	2.28236866	2.266793415	2.254343535	2.244155812

(continued)

(continued)

n	$\frac{n^*}{n}$									
	0.1	0.2	0.3	0.4	0.5	0.6	0.7	0.8	0.9	1
17	2.739234884	2.496302427	2.395277429	2.338286623	2.301270313	2.275146365	2.255662472	2.240543672	2.228455983	2.218562714
18	2.701333086	2.46561441	2.367538094	2.312187611	2.276224255	2.250835972	2.231895982	2.217196012	2.205440926	2.195818244
19	2.667052647	2.437937531	2.342565098	2.288720867	2.253725541	2.229014272	2.210575261	2.196261364	2.184813075	2.175440136
20	2.635852601	2.412812719	2.319931577	2.267476864	2.233375467	2.209289939	2.191314321	2.17735777	2.166193616	2.157052095
21	2.607298792	2.389873034	2.29929714	2.248129568	2.214856987	2.191352073	2.173806716	2.160182158	2.149282118	2.140355806
22	2.581038179	2.368821208	2.28038664	2.230415824	2.19791448	2.174950215	2.157805744	2.144490672	2.133836956	2.12511144
23	2.556780325	2.349413528	2.262974958	2.21412063	2.182339311	2.159880105	2.143110331	2.130084662	2.119661379	2.111123774
24	2.534283798	2.331448055	2.246875916	2.199066405	2.167959324	2.145973322	2.129554779	2.116800503	2.106593393	2.098232127
25	2.513346027	2.314755854	2.231934046	2.185105037	2.154631062	2.133089617	2.117001183	2.104502078	2.094498286	2.086302924

n	$\frac{n^*}{n}$									
	0.1	0.2	0.3	0.4	0.5	0.6	0.7	0.8	0.9	1
26	2.493795605	2.299194378	2.218018379	2.172111903	2.142233917	2.121111156	2.105333743	2.093075134	2.083263026	2.075224116
27	2.475486384	2.284642404	2.205017713	2.159981299	2.130665676	2.109938139	2.094454424	2.082422987	2.072791994	2.064900946
28	2.458292883	2.2709961	2.192836942	2.148622914	2.119839073	2.09948541	2.084279617	2.072463213	2.063003699	2.055252678
29	2.442106671	2.258165952	2.181394186	2.137959072	2.109679106	2.089679819	2.074737535	2.063125064	2.053828217	2.046210053
30	2.426833507	2.246074323	2.170618523	2.127922551	2.100120912	2.080458141	2.065766154	2.054347438	2.045205171	2.037713285
31	2.412391029	2.234653505	2.160448178	2.118454852	2.091108079	2.071765412	2.057311576	2.046077255	2.037082131	2.029710467
32	2.398706905	2.223844146	2.150829063	2.109504796	2.082591284	2.063553602	2.049326722	2.038268164	2.029413319	2.02215629
33	2.38571731	2.21359397	2.141713592	2.10102739	2.074527193	2.055780528	2.041770258	2.030879484	2.022158573	2.015011008
34	2.373365683	2.203856732	2.133059713	2.092982906	2.066877561	2.04840898	2.034605736	2.02387535	2.01528249	2.008239597
35	2.3616017	2.194591353	2.124830109	2.085336111	2.05960849	2.041405989	2.027800873	2.017224008	2.008753734	2.001811059
36	2.350380415	2.1857612	2.116991535	2.07805564	2.052689818	2.034742233	2.021326965	2.010897227	2.002544452	1.995697847
37	2.339661543	2.177333489	2.109514266	2.071113465	2.046094606	2.02839153	2.015158391	2.004869821	1.996629797	1.989875392
38	2.329408859	2.169278781	2.102371632	2.064484455	2.03979871	2.022330426	2.009272206	1.999119233	1.990987526	1.984321704
39	2.319589684	2.161570555	2.095539629	2.058146006	2.033780422	2.01653784	2.003647792	1.993625202	1.985597659	1.979017033
40	2.310174455	2.15418485	2.088996584	2.052077719	2.028020161	2.010994762	1.998266561	1.988369467	1.980442194	1.973943586
41	2.301136348	2.147099951	2.082722876	2.046261135	2.022500217	2.005684004	1.993111711	1.983335523	1.975504862	1.969085287
42	2.292450966	2.140296132	2.076700689	2.040679506	2.017204525	2.000589981	1.988168012	1.97850841	1.97077092	1.964427571
43	2.284096061	2.133755425	2.070913809	2.035317593	2.012118475	1.995698521	1.983421618	1.973874537	1.966226976	1.959957205
44	2.276051297	2.127461423	2.06534744	2.0301615	2.007228749	1.990996711	1.978859914	1.969421517	1.961860827	1.955662139
45	2.268298044	2.121399113	2.059988053	2.025198524	2.002523177	1.986472754	1.974471379	1.965138044	1.957661335	1.95153137

n	$\frac{n^*}{n}$									
	0.1	0.2	0.3	0.4	0.5	0.6	0.7	0.8	0.9	1
46	2.260819199	2.115554725	2.054823246	2.020417029	1.997990613	1.982115848	1.970245467	1.961013766	1.953618307	1.947554834
47	2.253599029	2.109915604	2.049841629	2.015806329	1.993620827	1.977916082	1.966172503	1.957039192	1.949722394	1.9437233
48	2.246623031	2.104470099	2.04503272	2.011356597	1.989404412	1.973864345	1.962243593	1.953205597	1.945965008	1.940028288
49	2.239877817	2.099207459	2.040386855	2.007058774	1.985332702	1.969952242	1.958450548	1.949504949	1.942338239	1.936461991
50	2.233350998	2.094117747	2.035895105	2.002904497	1.981397696	1.966172028	1.954785811	1.945929834	1.938834791	1.933017209
60	2.177925564	2.050996029	1.997896704	1.967799329	1.94817335	1.934276024	1.923881129	1.915794793	1.909315372	1.904001854
70	2.135539656	2.018138605	1.969011437	1.941159239	1.9229939	1.910128822	1.90050471	1.893017083	1.887016765	1.882095682
80	2.101785583	1.992045598	1.946114662	1.920070226	1.903081589	1.891048478	1.882045844	1.875041119	1.869427352	1.864822978
90	2.07409138	1.970684932	1.927397954	1.902849635	1.886835234	1.875491174	1.86700341	1.860398856	1.855105486	1.850763675
100	2.050842633	1.952786091	1.911733484	1.888449979	1.87325945	1.862498246	1.854446101	1.848180188	1.843157999	1.839038453
110	2.030968389	1.93750895	1.898377015	1.876181071	1.861699127	1.851439324	1.843761987	1.837787487	1.832998694	1.829070466
120	2.013726702	1.924273036	1.886815218	1.865567451	1.851703404	1.841880903	1.834530508	1.828810239	1.824225082	1.820463797
130	1.99858525	1.912662824	1.876681172	1.856269634	1.842950621	1.83351391	1.826451979	1.820956043	1.816550594	1.812936645
140	1.985150923	1.902372063	1.867704792	1.84803796	1.835204395	1.826111339	1.819306393	1.814010328	1.809765001	1.806282342
150	1.973126307	1.89316945	1.859682322	1.840684233	1.828286651	1.819502264	1.81292816	1.807811646	1.803710175	1.800345473
160	1.9622817	1.884876617	1.852456787	1.834063572	1.822060375	1.813555229	1.807189976	1.802235919	1.798264614	1.795006652
170	1.952436518	1.877353526	1.84590506	1.828062409	1.816418217	1.808167281	1.80199217	1.797186026	1.79333324	1.790172469
180	1.943446581	1.870488507	1.839929031	1.822590301	1.811274726	1.8032565	1.797255457	1.79258473	1.788840453	1.785768671
190	1.935195198	1.864191283	1.834449442	1.817574226	1.806560938	1.798756787	1.792915883	1.788369741	1.7847253	1.781735394
200	1.927586782	1.858387976	1.829401477	1.812954493	1.8022205	1.794614152	1.788921222	1.784490205	1.780938021	1.778023779

n	$\frac{n^*}{n}$									
	0.1	0.2	0.3	0.4	0.5	0.6	0.7	0.8	0.9	1
200	1.927586782	1.858387976	1.829401477	1.812954493	1.8022205	1.794614152	1.788921222	1.784490205	1.780938021	1.778023779
210	1.920542188	1.853017474	1.824731554	1.808681779	1.798206862	1.790784007	1.785228351	1.780904137	1.777437546	1.774593503
220	1.913995252	1.848028732	1.820394937	1.804714924	1.794481192	1.787229165	1.781801312	1.777576538	1.774189638	1.771410958
230	1.907890176	1.843378735	1.816353953	1.801019274	1.791010805	1.783918334	1.778609856	1.774477965	1.771165506	1.768447881
240	1.902179533	1.83903095	1.812576618	1.797565422	1.787767971	1.780824972	1.775628326	1.771583453	1.768340735	1.765680313
250	1.896822718	1.834954119	1.809035579	1.794328227	1.784728989	1.777926397	1.772834804	1.768871676	1.765694473	1.76308779

References

Filliben, J.J.: The probability plot correlation coefficient test for normality. Technometrics **17**, 111 (1975)

Ryan, T.A., Joiner, B.L.: Normal probability plots and tests for normality. Technical Report, Statistics Department, The Pennsylvania State University (1976). Available from Minitab Inc.

Vogel, R.M.: The probability plot correlation coefficient test for the normal, lognormal, and Gumbel distribution hypotheses. Water Resour. Res. **22**(4) (1986)

Glossary[1]

Alternative hypothesis In decision theory, any admissible hypothesis that is distinct from the *null hypothesis*.

Censored data Response data, such as fatigue endurance, are described as censored when its exact value is unknown but, for instance, it is known to fall within a certain range of values. Censored fatigue data are generally "right" censored, which means that the endurance is known to be greater than a particular value (typically because the test stops before failure actually occurs).

Characteristic curve/value A fatigue design (or characteristic) curve is established by adopting characteristic values that lie a certain number of standard deviations below the mean *S–N* curve (see Sect. 1.5).

Chi-square distribution The chi-square, or χ^2, distribution is the statistical distribution followed by the sum of squares of ν independent normal variates in standard form (i.e. having zero mean and standard deviation of one). It is useful in determining *confidence limits* for the standard deviation of a sample drawn from a *normal distribution* (see Sect. 1.7.3).

Confidence interval/level/limits Confidence limits are statistics derived from sample values, between which a population parameter under estimation will lie with some fixed probability *P%* (called the confidence level). The interval between the upper and lower confidence limits is called a confidence interval.

Degrees of freedom In regression analysis, the number of degrees of freedom f is equal to the sample size n minus the number of coefficients estimated by the *regression*. It is also used as a parameter of a number of distributions, including χ^2, F and *Student's t*.

Design curve See characteristic curve.

[1] This glossary describes statistical terms as they are used in this guide, as well as those that are commonly used in other guidance documents/standards on this subject. Kendall and Buckland [26] give more general definitions of these terms.

G. Parmentier et al., *Best Practice Guideline for Statistical Analyses of Fatigue Results*, IIW Collection, https://doi.org/10.1007/978-3-031-23570-2

Extreme value statistic The statistic given by the smallest (or largest) observation in a sample. An extreme value statistic is a particular type of *order statistic* (so the terms are often used interchangeably).

Gaussian distribution An alternative name for the *normal distribution.*

Hypothesis Conjecture to be tested by some statistical analysis.

Least squares method In *regression* analysis, a method of estimation in which the regression coefficients are estimated by minimising the sum of the squares of the deviations of the data points from the fitted regression line. In certain cases, the method is equivalent to the *maximum likelihood method* (see Sect. 1.4.3.1).

Linear regression See *regression.*

Log-normal distribution The distribution pertaining to the variate X when $\log X$ follows a *normal distribution.*

Maximum likelihood method A method of estimating parameters of a population (e.g. regression coefficients) as those values for which the likelihood of obtaining the observed data is maximised (see Sect. 1.4.3.1).

Normal distribution A symmetrical distribution that commonly arises as the sum of a large number of variates (e.g. measurement errors) having similar distributions to one another. For this reason, data are often assumed to follow a normal distribution in the absence of information to the contrary.

Null hypothesis In decision theory, the hypothesis under test.

Order statistics When a sample is arranged in ascending order of magnitude, the ordered values are called order statistics. The term can also refer, more specifically, to the extreme values of the sample.

Population The complete set from which a random sample is taken, e.g. the set of *S–N* data from all components of a given type, in the context of fatigue testing.

Prediction interval/limits Prediction limits are the limits between which a given proportion (typically 95%) of the population lies. The interval between the upper and lower prediction limits is called a prediction interval.

Random variable See *variate.*

Regression Process of estimating the coefficients of an equation for predicting a response y (such as $\log N$) in terms of certain independent variates (such as $\log S$). In the case of *linear regression*, the fitted equation is of the form $y = mx + c$.

Significance level In decision theory, the probability (typically set to 5%) that the null hypothesis will be incorrectly rejected when it is, in fact, true.

Standard deviation The most widely used measure of dispersion of a variate, equal to the square root of the variance.

Student's *t* distribution The t distribution is the statistical distribution of the ratio of a sample mean to a sample variance for samples from a *normal distribution* in standard form (i.e. having zero mean and standard deviation of one). It is useful in determining *confidence limits* for the mean of a small sample drawn from a *normal distribution.*

Tolerance limits Tolerance limits are values of a variate, following a given type of distribution, between which it is stated with confidence $\gamma\%$ that at least a proportion $P\%$ of the population will lie. This statement is made on the basis of a sample of n independent observations. A tolerance limit can thus regard to be a confidence limit on a confidence limit.

Variance The mean of the squared deviations of a variate from its arithmetic mean.

Variate A quantity (also called random variable) that may take any of the values of a specified set with a specified relative frequency or probability.